Trolleybus Revolution - UK

The Retrofit Hybrid Revolution

Terence P. O'Halloran
F.C.I.I., B.Sc. (Hons)

Chartered Financial Planner

Trolleybus Revolution - UK

A catalogue record for this book is available from the British Library.
Design and layout by Life Publications Ltd, Lincoln
Front cover artwork by Life Publications Limited
Initial editing by David Chick
Art illustrations Linda O'Halloran
Graphics and layout by Life Publications Limited
All in conjunction with the author

Life Publications Limited
Tiptree
Stainton by Langworth
Lincoln
Lincolnshire
LN3 5BL

www.lifepublications.co.uk
e-mail: tpo@lifepublications.co.uk
Tel: +44 (0) 7798 678705

ISBN 978-1-80227-179-9 (Paperback)
ISBN 978-1-80227-180-5 (eBook)

Contents

Preface

An Environmental Coup

Are you part of that group concerned about overhead wires for 'in motion' electric bus charging being unsightly, whilst also being in the fraternity that discards cans, bottles and product wrappings in the street, countryside and/or the beach?

Or are you, perhaps, part of an 'extinction rebellion' or a similar environmentally aware body that believe it is a great thing to glue yourself to a building or smash windows or have meetings of thousands of people leaving rubbish for 'tens of people' to be employed to clear the streets after you?

No, those are not politically-inspired observatory questions! They are questions born of observation from a person that lives in a small hamlet that clears away fifteen cans, six plastic bottles, a glass bottle and sundry waste on average every week from three miles of grass verge. All because people are environmentally blind or simply cannot be bothered to take their rubbish home.

Perhaps, instead of any of the above, you, or people you know, are part of the magic roundabout brigade that circumnavigates McDonald's or Pizza Hut or whatever your favourite instant food outlet is - to then throw the empty cartons and wrappings from your vehicle on the way home once you have finished your meal? However, one ponders, although you have never perhaps experienced them; the proposed overhead wires to be erected in the middle of a town or city will be offensive. Gosh! That is unthinkable. How unsightly.

Yet, how practical in terms of providing valuable electric current to a means of taking passengers from A to B, totally free of vibration, noise, and environmentally unfriendly excretions. Surely the overhead wiring is worth the clean air experience, just as our taking our litter home is perhaps considered inconvenient but leads to a cleaner environment.

Alternatively, and contrary to the above, you could be the kind of person that gets together with a couple of friends who are handy with breeze blocks and a bit of cement ('gobo' as those in the trade might call it). You all descend on Civic Centre such as Scunthorpe whilst their full council meeting is in progress and build a wall, a proper wall, with three sides so that it cannot be pushed over, across the entranceway to the building. You build it on a plastic sheet so that, when it is dismantled a few hours later, after your point has been made, there is not a speck of dust left, let alone cement residue or brick dust, to undermine your message.

The point of the exercise? In the mid-1970s, communication with Scunthorpe Civic Council was akin to "talking to a brick wall." The tradespeople of the area breeze-blocked the whole council into their building to 'cement' the point they wished to make, which was, "listen to us." It worked. The point was made, and the result was the application of common sense and democracy. Prejudice can often be blindingly obvious and turn a deaf ear.

Or perhaps you might need to be extreme and take 12 wheelbarrows 150 miles from Langworth in Lincolnshire and venture south to Victoria Street in the heart of London as happened in March 2008. The objective was 'The Department of Business and Enterprise,' where a group of sixteen agitated and frustrated individuals built a breeze-block wall (transported through the streets of London in a column of those eight wheelbarrows). The wall across the Department's front entrance was accompanied by a large sign "trying to talk to this government department about sub-post office closures is like talking to a brick wall." The wall made their point. The cleanliness afterwards gave credence to that point.

That was a demonstration against the closure of sub-post offices. The vendetta that the Post Office hierarchy had against sub-postmasters founded, as it now transpires, twelve years and millions of pounds of wasted public resources, on a misconception. That misconception centred upon the Post Office 'Horizon' computer system malfunction which cast a shadow on all sub-post office operatives. The true position has now come to light in the courts. The message is clear: "Do not let pre- judgement or misconceptions colour your view."

On that sunny day in 2008, there was a message to take note of. It was ignored, but the point was made. No mess, no breaches of the peace, no violence - just a simple message. Hopefully, that may be your style. It is the style of this book: a simple message based on factual evidence that deserves your consideration.

We can all get emotional about the view of electric wires above the street, hampering our view of the sky. What of the positive environmental effect of that inconvenience? What of the improvement in the lives of the people who abide in the vicinity of a busy thoroughfare when the vehicles using it to carry passengers change from diesel engine drive to electric power in an economical and ergonomic way? Isn't that worth the wires?

Which category do you fall into; ban the wires and drop the rubbish, or put up with the wires and save the planet? It is a simple enough metaphoric question. A survey of 1,077 people from the UK's 68,000,000 population will fail to state who you are. Only you can truly do that. The statistics quoted in the following chapters are far from speculative statistics. The case studies are based upon facts.

Here is the argument for the wires, also known as the catenary, or the 'In Motion' Charging (IMC) facility for electric buses of all types. In the modest discomfort to your emotions that it might cause, there is a compelling argument for the installation of the catenary system to help save our planet that requires urgent implementation. It is compelling in every facet, bar that minor visual impediment. Consider it seriously and hopefully recognise the good economic, environmental, and practical reasons to support its implementation. Time is running out. The management of the Post Office lived in denial for 20 years; please avoid the 'brick wall' syndrome that impaired their objectivity.

The purpose of this publication is to examine and put forward the case for re-introducing something akin to, but intrinsically different from, the conventional trolleybus network system utilising an overhead wire power connecting facility (a catenary) in the United Kingdom. The following discourse more than justifies that cause in financial, environmental, and public health interest terms. The provision of an 'in motion (electrical) charging system' for public service vehicles (buses) is quicker to install and economically better value for money than many if not all practical alternatives. It is not just the author that comes to that conclusion.

The author is a septuagenarian and, by virtue of that fact alone, understands the utility of the trolley bus in service first-hand. He is also an engineer, having served an apprenticeship with the Royal Air Force at the RAF'S No 1 School of Technical Training Halton before completing a 12-year service engagement in the Royal Air Force.

Much has been written about the trolleybus. This publication starts, quite rightly, with a relatively detailed history of the progression of public transport over the years. It covers some of the perceptions that have led to misconceptions that resulted in the trolleybus, in the United Kingdom at least, being side-lined in favour of other forms of transport propulsion - predominantly diesel power. The error of that judgement has now come home to roost, as the following chapters outline and confirm.

Most published works, particularly from the European Union, argue for the instigation of a new trolleybus era on the basis of new vehicles, including a new infrastructure using mass-transit public service vehicles. (1) The new wave of 'bendy bus' articulated vehicles are almost 26.25 yards long (24 metres). The 2013 trolleybus report (1) from the EU uses those mass-transit vehicles as their example of what the universal future transport design should be.

The author, on the other hand, argues that the existing fleet of hybrid vehicles can more readily be made the recipients of overhead power provision, which would then fit into the current street layouts. The transition would utilise a revised infrastructure that can provide more than one useful outcome for society by way of facilitating a power supply to the retrofitted buses: In Motion Charging. Put another way, the provision of an electrical charging/power supply whilst the recipient vehicle is in motion fulfilling its design function of transporting passengers to their destinations.

There are some 44,000 trolleybuses operating in the world today (2021): none in the United Kingdom (except at Sandtoft Trolleybus Museum and similar facilities).(24) Yet the trolleybus was developed in the UK and is proven to be the most economical, environmentally friendly, quiet and safe vehicle that all analysis confirms is the best of the best in every respect, except perhaps, for the wires. There is no more economically and environmentally acceptable solution which the EU figures more than ratify. (1)

Whilst the 2013 report (1) is an extraordinarily detailed insight, and all of it pertains to what is judged best for Europe, and by example, Britain; the footprint and manoeuvrability of the lengthy, inflexible in operation, 'power trains' would not do well on the British urban road system. Even the single-deck, single articulated hybrid buses, in many circumstances, have too large a footprint to be the best that they can be.

However, double-decker hybrid buses can carry 65-78 passengers in normal circumstances with ease. They do not interfere unduly with other road traffic or demand mass demolition of existing buildings to create an infrastructure that accepts the need for huge turning circles that accompanies the 'power trains' of the EU report's incredible length and suggests something more acceptable. The report itself is a valuable assembly of comparative data, which more than sustains the argument for a more durable UK alternative strategy utilising conventional sized vehicles for most applications.

The retrofitting of the current hybrid fleet of buses does two things in a short space of time:

A. The capital cost of providing the vehicles fit to run on the system is often less than 10% of the capital cost of comparable new vehicles.

B. The retrofitting extends the life of the recipient hybrid vehicle, thus amortising the original cost of the host vehicle over a further 10- to 15-year lifetime period.

The foregoing always assumes, of course, that capital is spent on designing and implementing the installation of the appropriate catenary (overhead wire power system) for the retrofitted vehicle to draw power from (In Motion Charging).

The rejuvenation of the fleet of some 38,200 vehicles currently in service (9,300 in London)(7) allows for experimentation and development to take place. The vehicles working 'in service' are earning a living and concurrently providing a public service, once again saving unnecessary capital cost when capital is in restricted supply.

Many northern cities in the United Kingdom had a thriving trolleybus system from the 1930s through to the 1960s. They thrived on their efficiency, low-cost running, environmental benefit and, of course, improved health for the local population. This publication is devoted to a return to those healthier days of public transport provision in urban and city locations with distinct suburban and rural implications.

There is a lot of relevant detail in the 2013"*Take-Up Guide for The Replacement of Urban Diesel Buses by Trolleybuses,*(1) a three hundred plus page EU sponsored paper, as already mentioned. However, the basic tenets of that paper are to compare the operating costs and amortised infrastructure costs with the operation of diesel and other source powered Public Service Vehicles (PSVs) using a more mass-transit facility by way of articulated trolleybuses as a diesel alternative. There are circumstances where one could be convinced of the mass transit strategy, but is it universally appropriate?

That principle would not work as cost-effectively in the United Kingdom because the UK urban street configuration will not allow it to do so without major structural changes. This process would involve the possible demolition of historic buildings and/or radical redesigning of an urban street layout against the public interest and necessity. Additionally, it would be time-consuming in obtaining planning consent and infrastructure costs. That may well be unacceptable socially, but surely the expense and loss of facility in city centres which that would involve would be a more pertinent objection. (See the Leeds City submission (6))

The Leipzig and Swiss case studies simply do not apply to the United Kingdom in the specific form that they are presented. The general observations are interesting and pertinent by way of analysis. They are worth taking note of as they support the EU and the author's viewpoint in analytical scientific terms.

As will be seen later, the cost-effective nature of the retrofitted hybrid fleet utilising the proposed auto-connect and disconnect system facilitating the In Motion Charging system drives an unassailable economic argument compared to the current hybrid bus alternative in all its formats.

Foreword

A Personal Experience with Public Service Vehicles

The author's personal experiences involving the utility of public service vehicles centre upon the London red bus fleet, including trolleybuses, operated by London Transport Executive together with their green 'Country' buses and the ubiquitous Green Line coach network along with Southern Vectis and other regional operators dating back to his age six or seven years. The 1955-1965 period. He writes:

We had lost my mum when I was aged four. Dad was bringing me up on his own as a self-employed tailor in Bedfont, Middlesex, west of London and just south of London Airport (Heathrow), about a mile off the main runway and 15 miles from Hyde Park Corner.

The historical memories are London orientated but are far from confined to that city's transport network.

Feltham was a mile away to the south of Bedfont on the main railway line west of London, forming part of the Southern Region network which provided trains every half hour travelling east into London or in a westerly direction out past Windsor into the countryside and southwest to Reading and beyond.

The main A30 road ran past our home. That was on Staines Road, Bedfont, which was well serviced by public transport. Every half hour, a 701 or alternately a 702 Green Line coach, running from Ascot and Sunningdale respectively, came past on their way into London Victoria and out to Gravesend and Sevenoaks. These were Green Line buses (single-deck circa thirty plus seater limited-stop coaches), which I became very well acquainted with following a little incident where I locked myself out of our maisonette home. The Green Line service to Hyde Park, followed by a number 9 or 73 double-decker red bus, would get me to dad's workshop just off Regent Street. The journey home was the reverse journey or by train to Feltham followed by a mile walk.

A 'red' AEC bus, the 116, from Hounslow's central bus terminus also ran past our house on its way to Staines. Again, this became a familiar means of transport for me from a very young age: to and from Hounslow, to the larger shops as well as the ABC Minors Club experience at the Regal cinema on a Saturday morning. Also, on a Sunday to go to church, incorporating a mile walk from Hounslow Heath and the George IV pub bus stop across to Hounslow West near the tube station. Public transport was an essential part of many people's lives, including dad and me.

Dad originally worked from home on pretty much a six-day week with his customers coming to our home for fittings. I was the soul of discretion when I was at home, but of course, most of the time, I would be at school. I initially travelled to infant and junior school from the age of five or six on the Bedford 29-seater Duple coach run by Bedfont Transport. I can still picture the apple green livery with the pale-yellow stripe down the side and those curvy lines of that well-defined Bedford Duple built vehicle. That facility lasted until the unfortunate day that the 'bus' caught fire. After that, I walked the 1 ¼ miles to school through the adjacent housing estate.

Dad and I would find entertainment by catching the 116-diesel powered AEC bus to Staines, and there we would transfer to a London Transport green bus, the 'Country Bus' service, 441, that travelled via Runnymede to Windsor. What great days out they were. What a fabulous viewpoint from the top deck of those buses, with their unique front entrance and staircase, providing a panoramic view of the countryside. I would swim in the River Thames at Runnymede or walk the Royal Mile to Windsor Castle and more, courtesy of that excellent bus service.

My father's work moved from home in my very early teens (circa 1953, because I remember the coronation street party) to employment as a repairing tailor with a dry cleaner opposite the trolleybus terminus outside the Duke of Wellington Pub west of Hounslow. However, he also picked up a lot of work from Hounslow Army Barracks, where a variety of regiments were represented. Made-to-measure mess dress garments were always in demand. I became accustomed in my teens to being a 'tailors dummy,' a model, to try on mess dress ensembles. Dad's reputation spread, and he was duly employed in the 'West End' of London off Piccadilly.

It was an interesting life, and most of my time after school was spent at home until one fateful evening when I locked myself out as I went bus gazing, looking for my dad to come home. Invariably he would be on the 10:30 pm or 11 o'clock Green Line bus from London.

Unfortunately, the next-door neighbour in our block of four maisonettes unhelpfully reported the fact that I was 'home alone' to the police, and they, in turn, warned my dad on his eventual return that "things had to change." From that point on, I left school at 3:30 and, by 4:30, was on the Green Line bus, single-decker, fast track transport, into central London. I got to know the buses and their crews very well.

At weekends, as the evenings 'drew in,' dad and I would invariably go on a Saturday evening to the Chiswick Empire, which entailed catching the 116 from near to our home to the Duke Wellington bus stop, which was opposite the terminus for the 657 trolleybuses to which we would transfer. The trolleybus would take us to the Chiswick Empire, Turnham Green, for our weekly entertainment. That happened for most of our winters prior to me joining the Royal Air Force as an engineering apprentice in January 1960, aged 15.

From age 6 to those early teenage years, every other weekend from late spring to early autumn, dad and I would go on an excursion train, or by train from Feltham station to Waterloo and then down to Portsmouth, normally hauled by a steam engine, to make our way to the Isle of Wight.

The Isle of Wight *Southern Vectis* bus company ran Bristol double-decker buses powered by Gardner four-cylinder diesel engines. I would delight in using those buses to go to Sandown, Shanklin, Ventnor and other beauty spots on the top deck, preferably in the front window seats. They were a compact vehicle ideally suited to the narrow country roads of the island. Otherwise, we were on the beach at Ryde, which was our place of arrival on the ferry from Portsmouth. Those Bristol bus journeys occurred later in life as well; Wendover to Aylesbury from RAF Halton, in Ipswich, from Swindon to Wroughton RAF hospital and also the aircraft repair unit: so many places creating so many memories.

I can still recall the comfort of those buses and the scenery passing by viewed from those front seats upstairs to enhance our appreciation of the surrounding countryside. We could capture the magnificent features that the Isle of Wight had to offer. The double-decker bus was part of the experience. That experience was reflected 30 years later with my own children as we took our caravan across to Ningwood on the island and used those same, or so it seemed, Bristol buses to go to Freshwater, Alum Bay or Newport or around the island, ignoring the fact that we had a car to use.

The double-decker bus was just such a tremendous experience. As the driver of a car, I saw little or nothing but hedgerows. As a passenger on the top deck of a bus, I saw and enjoyed the beauty of everything we passed.

The experience that I had then and now has never left me. Nor has it changed my view of the fact that the double-decker bus, and indeed those trolleybuses from the Duke of Wellington Pub to the Chiswick Empire (and 667 to Shepherd's Bush Empire; oh yes, and the Hammersmith Empire), were indeed an experience in their own right. They were not just a method of getting from A to B; they had a unique integral entertainment value, varying as it did on every occasion, just as the stage shows at the various theatres varied, with differing music hall acts performing each week. Other people can experience that level of enjoyment today if only they are attuned to the possibilities.

The trolleybus routes were in urban London and other cities: clean, silent, vibration and emission-free. Everyone could breathe fresh air - even, or more correctly, especially in the cities.

Chapter 1.

Moving People from Place to Place

Can you answer a question, please? In what year was the first hybrid public service vehicle created and operated?

The year was 1879. Yes, over 140 years ago. In the UK. More of that later.

There is so much more to public transportation than just the wheels the transport runs on and the power it depends upon. Should we really be going back to natural power and horse-drawn vehicles that entail journeys that take days or maybe months instead of hours and maybe days?

The progression of human communities in terms of getting from one place to another has changed immeasurably over the millennia that our species has been evident on Earth. The planet itself has experienced huge changes in its geography, geology, and topography through the intervention of human endeavour and inventiveness. The changes have been incremental and detailed in terms of the structure of the planet; however, those changes have also been lifestyle changing for the inhabitants in extraordinary ways. We forget those changes at our peril, perhaps.

We can now be on the other side of the earth, England to Australia, within 24 hours of leaving the United Kingdom. An individual can leave Lincolnshire driving a car, and nine hours later, be in the north of Scotland. If an individual or group makes their way to Humberside Airport, they can do the same journey to Scotland's major cities in just over an hour and have a cup of tea and a biscuit while that is taking place. The journey would take four to six hours by train. These are huge improvements on the journey times that were available to the enterprising traveller in the year 1500.

Even in 1800, the journey times from Lincolnshire to Scotland would run into days. Is the environment any better or worse off in 2020 compared to 1520, or even 1820? Indeed, what constitutes the environment in travel and human health, or planetary health terms?

These surely are the questions we must ask ourselves as a central focus of our deliberations. Has the environmental impact of human progression been more or less detrimental due to the environmental effects resulting from the progression of transportation in overall terms? Are there elements of the transportation system that need to be taken into account in order to truly assess and quantify the advantages or disadvantages to the world as a whole coupled inexorably, as it is, with the advancement of technology and the underlying change that has taken place in transporting not only people but goods, from one place to another?

There may well be an overarching assumption that the main function of the public transport system is to take people from home to work and return them to their homes in the evening. Is that really the case?

The New Routemaster's introduction into London in 2013 blew that 'utility' concept out of the water. The bus's introduction reversed the concept of utility by making travel on public transport, at least above ground, into a pleasure and, to some, a real pursuit. For many people, public transport provides an entertainment element, an experience like no other. It should be an enjoyment personified, particularly for those who would otherwise be driving. By making public service vehicles attractive, will people abandon the car and its frustrations for some if not all of their journeys? Who knows? Some indicators suggest that it will.

The story behind the New Routemaster and its design is the pure study of humanity and its overriding need to want to accept something for its own attributes rather than be forced to accept substandard alternatives. Of course, there were huge debates but the great merit of this new form of public transportation, as it was developed in 2011, was that the design decisions were based upon its technological background and capability, long before its aesthetic qualities were considered. The marriage of the two, technology and persona, was a masterclass in engineering finesse for a public service vehicle. So, what was the problem? Some would immediately state: batteries.

The intention of an auto-connect facility, far from changing that technological prowess and aesthetic excellence, is to enhance it from within the vehicle itself. This maintains its persona and thereby its attraction to the public, many of whom do desire to travel on a bus rather than have to travel on it.

To ride on a bus for pleasure rather than bland 'need' and take the experience as something such that "travelling on this double-decker is an enhanced experience altogether" must be the desire of more than a few potential passengers. Even with 44,000, mainly single-deck trolleybuses operating the world over, and tens of thousands of single-decker buses (Department for Transport (DfT) designated 'coaches') along with them, the double-decker public service vehicle (DfT designated 'buses')(7) has an added dimension that is undeniably attractive to a good number of fare-paying passengers.

The difficulties experienced by not just the 'New Routemaster' but also by the 'Dennis Trident' since 1999, the 'Enviro 400' since 2008 and of course their predecessors, the 'MCW Metrobus' format, stem from the fact that diesel is now deemed to be 'dirty.' The Euro VI engine is a move in the right direction; however, is it enough? It has to be a step in the right direction, but can it be improved upon? The 'Wrightbus Pulsar' format using hydrogen as a fuel also has merit; however, it is technically in the 'development stage' of its useful life. What in the meantime? What about NOW?

What is required is a mechanism whereby the existing fleet, as an entity, can be modified into a more ergonomically acceptable and environmentally friendly format without expensive and time-consuming prolonged development costs. This is achieved by utilising the extensive development incorporated in the production of the 'New Routemaster' and similar hybrid design criteria to enable the existing fleet to be retrofitted and modified (as mobile 'in service' testbeds) for future development as a 'Powertrain' alternative with all the benefits that that concept delivers, and more.

The lessons of the Coronavirus restrictions take that argument even further where mass transit, standing room only, long-distance travel is encountered.

The public affection for the New Routemaster's 'look' as married to the hybrid concept is not an accident. That public affection is the culmination of a process that created a design in technological advancement and latterly introduced and refined the vehicle's aesthetic appeal, which made the bus a pleasure to ride on (reportedly with the notable exception of the mid-summer sauna effect of an inferior air conditioning system). That excellence now requires to be enhanced and taken forward to its next step: to overcome the problems of battery duration and diesel exhaust deficits (and passenger overheating in summer). The innovative interjection of retrofitting and in motion battery charging will counter this in spades very simply and relatively inexpensively.

It is time to use history, exactly as the hybrid motor as evidenced from over a century ago, using old technology in a modern environment, to enhance every aspect of public travel by bus.

Despite the foregoing narrative, the underlying principle is far from restricted to the New Routemaster genre, which is merely used as an example. All hybrid vehicles, as well as battery-powered vehicles, can, and should, benefit from retrofitting, as the following examination of their collective Achilles heels will demonstrate.

Chapter 2.

The Stagecoach Era is not just a Reference to a Transport Company with a Scottish Heritage

It was a Scottish initiative that set the scene for stagecoach transportation by way of people from Edinburgh being able to travel to Leith in 1610. If the reader compares that humble beginning and progresses the public transport facilities through to 1673, when it took eight days to go from London to Exeter, and then considers a move further forward in time to when the first stagecoach company was formed undertaking the journey from York to London, the journey times and availability of public transport improved tremendously.

It should be noted that this advancement was based in York, <u>not</u> in London. Other routes were initiated. However, the greatest and most significant improvement was in the infrastructure, which improved measurably.

Perhaps one of the first things to be considered is the fact that the fuel for 'the engine' involved a high cost. Lots of it was required in order to feed not only the 2 to 4 horses required to pull a stagecoach arriving at a staging post. There was also an overhead cost involved to feed the replacement horses prior to departure on the next 'stage' of the journey. The system involved a downtime when the motive power was not fulfilling its reason for being there but recharging its metaphoric batteries. The horses had to have a break; therefore, significant 'extras' were required to fulfil the passenger travel timetable.

A further consideration was the fact that the 'service station,' as we would know it today, was essential to the efficiency of the system. A facility was required for 'refuelling'/replenishing the horses, feeding them and allowing the animals' physical recovery, which took a great deal of time and the capital cost of the facility.

If the journey had to continue, a facility to exchange the horses was necessary, even at an overnight inn. The 'stages' were required every 10 to 15 miles of the journey. This is surely the precursor to today's motorway service area and, as such, requires understanding, particularly with all-electric battery-powered vehicles.

Those 'staging posts' where the horses would be changed gave their name to the vehicle making the journey, the 'stage' coach. The requirement at each 'stage' meant that stabling had to exist with the attendant services that not only arose from feeding the animals but clearing up what came out of the other end of them. It is to be left to the reader to assess the environmental impact of that particular facet of the public transport system. The impact in an urban environment bares contemplation as the number of vehicles and motive power increases proportionately with the active population.

Accommodation might also be needed for those travelling. Hospitality by way of food, drink and toileting would also perhaps be a prerequisite at those stopping points. Once established, and with a ready customer throughput, shopping facilities might be added, and staff required to attend them. A village and then a town could be a further consequence of regular traffic and invariably was.

This was, of course, a positive side to this evolutionary advancement. In the longer term, coaching houses were established, which then became centres for the aforementioned hospitality but also begat the formation of villages and towns that grew around them. Using Lincolnshire as an example, the city of Lincoln would perhaps be a journey's starting point with The White Hart Hotel possibly as an inn on the journey to York. The carriages would progress north 12 miles to Spital in the Street, previously a resting place for Roman legions on their way north on what is now the A15. The horses would be changed at Cromwell House.

At Spital [the name means resting place or more specifically a shelter], the horses would be changed, and the refreshed 'engine,' comprising of a fresh pair, or more, of trusty steeds, would take the carriage and its contents to the Angel Hotel in Brigg, a further 13 miles north. And so the journey would progress with that first 25 miles being covered in as little as six hours.

The 'roads' were poor dirt tracks. The vehicle wheels were thin, the axles unsprung and rigid, ideally destined to settle into the mud in bad weather. Despite the impediments, the carriage of passengers and their luggage was relatively more comfortable than the alternative open cart, personal mount, or walking. It is that comparative advancement that is so pertinent in this thesis.

The journey time from Manchester to London, using what was then termed 'the flying coach,' would be four and a half days in 1754. A mere three years later, the Liverpool to London stagecoaches, which had by then been upgraded with steel sprung suspension on improved highways, achieved 8 mph on their journey and shortened the time spent on that journey to an incredible three days. Is it pertinent in consideration of the 'advancement' of travelling facilities to ask questions such as:

a) How many horses were necessary to complete a journey?
b) How much stabling was required, for how many horses?
c) What was the effect on the environment of the facility?
d) Was there any adverse or positive effect on public health and living standards in the towns and cities connected to the service?
e) What was the cost, in current socio-economic terms, of providing the service?
f) Was that cost a capital expenditure or out of revenue?

In answering those questions, we perhaps must think that by the start of the 19th century, in 1800, there were 42 stagecoach routes in Great Britain.

The journey from Cambridge to London took two days in 1750. However, due to coach design and, more importantly, metalled road construction by 1820, that same journey had been reduced to seven hours. Speeds of 12 miles an hour were being achieved with horse-drawn carriages on decent 'metalled' roads. Capital expenditure on the infrastructure, amortised over time, was essential to progress being achieved. Society, as a whole, appears to have been the beneficiary as well as the paymaster.

It was a mere thirty years later that those significant achievements on long haul stagecoach-based journeys were dwarfed.

The advent of the railway network changed the dynamic pertaining to travelling long distances for many years. Consider the infrastructure cost and the amortising principal that was necessary for making the decision to move forward, to improve the facility, to make the capital investment for the long term gain.

However, in spite of the achievements regarding the influence of the railways, what was happening in the cities and large towns at that time was, if anything, contrary to that external progress.

Horse-drawn buses had been developed alongside the hansom cab, and smaller horse-drawn public service vehicles entered service. Commercial horse-drawn vehicles were developed and became established to deliver milk, coal, goods and chattels. The era of commercial haulage and thereby public transport had started with a vengeance. The environmental effect of all those horse-drawn vehicles on narrow streets can only be imagined.

***2.1** 300 cubic feet of HORSE MANURE produced by three horses over four months – visible emissions!*

However, the roses grew very well in the gardens adjacent to those busy streets. The knacker's yards also thrived, enthusiastically producing glue, tanning leather and other by-products of the horse-drawn vehicle age. There was a resistance to change, as there inevitably is.

Chapter 3.

The Advent of the Rail Network

The Stockton to Darlington railway came into being in 1825, the first railway to be available for public use. Within five years, the Liverpool to Manchester railway had opened, giving birth to a major development throughout Great Britain: a means of distance transportation that would herald the demise of the stagecoach outside of urban areas as the main means of passenger movement between large or distant conurbations. The infrastructure cost was immense as a capital investment. The success of the ventures by private enterprise can be seen today; however, the capital cost was always acknowledged to be assessed and therefore amortised over a prolonged period. There have been notable changes over the period of development from those rudimentary beginnings. What is the current life expectancy of a carriageway, a road, a railway track? What was it then?

The steam engine was to be a mechanism that changed not only the face of railways but also shipping, as did the introduction of steel for the tracks that the railway trains required to run on. Also, the steam engines to be mounted in the holds of metal floating vessels were made from that newly mass-produced material made from iron ore. Steel would replace wood in so many structures over the next two decades and beyond. That progressive step saved the remaining hardwood forests from total obliteration. The source of manufacturing material moved underground, literally.

The Luddites feared mass unemployment. But what of the environmentalists?

Even in agriculture, where steam locomotion hauled ploughs from one end of a field to the other, replacing the horse-drawn plough, there was suspicion and fear of change. Steam engine-powered threshing machines were displacing the previous requirement for large armies of agricultural workers who carried out those tasks by hand prior to the 'new' innovation. Days were saved, and jobs lost. Agricultural workers moved in large numbers to the growing factory-based conurbations and new work incorporating new working practices.

Road construction became less labour intensive as steam-driven road rollers and lorries supplanted the smaller capacity capability of the human and horse-powered predecessors. The requirement for roads grew, and therefore work and earning power was available. Factory employers provided large estates of housing to facilitate the population move from rural to urban areas. Large areas of terraced housing were erected to house the influx of workers to growing industrial centres incorporating modern facilities that today appear arcane. The population dynamic changed dramatically as mass production and new sources of power - energy - took over the population's lives and livelihoods.

The mills of the north progressed from the click-clack of hand-operated looms to the more aggressive and constantly noisy steam-propelled mass production units in the production of cloth, millinery, hardware and associated products. Steel mills grew in number, as did coal and iron ore mines. Huge change was sweeping through society and communities empowered by the by-product of the change in public transportation in almost every environment. Progress emanated from the north. Wealth, however, dispersed over time to the south.

The bicycle was created and started to make its presence felt in respect of the working individual. The bicycle was relatively affordable for working folk. The bicycle became even cheaper as mass production developed. The 'bike' was easily stored in the owners' terraced accommodation, in the cycle racks that swiftly appeared at factories and on school premises. Public thoroughfares improved dramatically and at pace. The change created employment at all levels. Skills improved.

Transport, agriculture, factory units; all became heavily influenced by the steam engine, with one notable exception. That exception was in the transportation of people within cities and towns.

The horse-drawn vehicle remained a major mechanism for a significant period. The steam engine was all-powerful in so many respects, but to bring it to a commercially viable mechanism for mass acceptance in urban transportation proved to be a step too far for a fast developing and growing marketplace. Urban steam power was left behind by other developments. Steam power was heavy, noisy and cumbersome by comparison to other innovations where provincial travel was concerned.

It was to be the petrol engine developed by Daimler and Maybach in Germany that would prove to be the innovation that would change so much with regard to the application of power and the environmental cleanliness of transportation. However, it was not employed as hindsight would expect.

The petrol engine took centre stage, and the horse-drawn bus and hansom cab disappeared. Internal combustion engines in one form or another had been developed and researched as far back as the 1790s, but it was the four-stroke design of that German duo and the two-stroke internal combustion engine of Karl Benz that went into 'mass' production in the late 1870s. Fifteen years later, Rudolf Diesel developed his compression engine, which still bears his title.

By this time (1900), the evolution of power and transportation, particularly electricity, harnessed by Michael Faraday in England (1831) and Nikola Tesla, had started to make its mark. Electricity had overtaken gas as a power source which had overtaken candle power for domestic lighting a century before.

Whilst this publication is primarily focused upon the evolution of energy in providing for transportation, those three illuminating elements in the domestic environment, along with street and home lighting in particular, bear thinking about in respect of their environmental effect and, of course, their application in society as a whole. The world as an entity has been influenced by the relatively speedy move from one power source to another in the developing world.

It is, however, worth calling to mind the fact that in 2009 the International Energy Agency estimated that 585.2 million people in the sub-Saharan area of Africa remained without electric light. This is perhaps a fact which is possibly difficult to absorb as we in our developed world take electrical power for granted. Thomas Edison's light bulb of the 1880s was a transforming, safe means of light that still leaves areas of the world in the dark in 2020. How can that be?

There will always be the sceptic:

"When the Paris exhibition closes, electric light will close with it, and no more will be heard of it." Professor Erasmus Wilson, University of Oxford 1848

Chapter 4

The Horse-Drawn Bus

In 1829, George Shillibeer started the first omnibus service in London.(7) This was from 'The Stingle' pub in Paddington, now part of Greater London, to the Bank of England in the City of London. His vehicles travelled via what is now the Marylebone Road, around the perimeter of the City of London north of Clerkenwell and down into the City itself. The circuitous route that it was obliged to follow was due to Hackney carriages holding 'priority' over the movement of people within those inner-city limits, with their owners and operators presenting a strong lobby with the authorities.

There were just four journeys per day in each direction hauled by three horses pulling a sprung chassis carriage which would take 22 passengers. It proved a tight squeeze because those were busy buses. Numbered routes were apparently introduced in 1906.

Shillibeer was a coachmaker who started his entrepreneurial excursion on a bus service in Paris, but the Paris vehicles were small compared to his English enterprise. The first real London buses, by contrast, could accommodate up to 22 people, including a lower deck giving cover against the weather. There were apparently two tariffs for the 3-mile journey: sixpence or a shilling. Only the relatively well-off could afford to use the service, and they did. Excitement and entertainment value had to be a factor implicit in customer (passenger) satisfaction.

There is a road in the Marylebone area of today's London known as Shillibeer Place in George's honour; it marks the place where his horses and coaches were stabled. His bus services were always the subject of controversy, particularly with the Hackney Cab drivers who resented that particular form of public service vehicle because it took away their customers. Shillibeer was competition. The London black cabs and Uber conflict of today comes to mind by way of its similarity. Change is uncomfortable for vested interested groups, often unnecessarily so. Fear of the unknown?

It is perhaps interesting to note that various types of steam-powered vehicles were trialled in the 1820s and 1830s to perform much the same omnibus service. However, they were restricted in their advancement, largely because of penal taxes and licence fees that the hansom cab operators agitated for with the government on the grounds that those motorised vehicles were a danger to the public.

Most of the history that relates to horse-drawn vehicles indicates that horse-drawn transport lasted well into the Victorian era. The fact that their supporting infrastructure was in place might underpin that circumstance.

Steam carriages were introduced notably by Goldsworthy Gurney, an inventor. Goldsworthy Gurney was one of several individuals that succeeded in getting steam-driven vehicles onto the streets and used by the public. The advantages over horse-drawn vehicles were obvious; the omnibus was faster, and they had more power to negotiate hills. Additionally, of course, the detriment of the huge amount of horse manure was avoided, which was a real and major problem in large conurbations. What did one do with tons of manure per day?

Gurney's carriage works were close to Regents Park. He organised increasingly ambitious trips in his vehicles, going as far as Barnett or Stanmore at speeds of up to 20 mph. A later model of his went from London to Bath and back. The reliability of these alternative means of transport compared to the horse-drawn vehicles of the time was a stark contrast and became more acceptable as time and experience moved on. These were all hybrid electrically motivated vehicles with the onboard electricity generated by a steam boiler.

This was the advent of the hybrid vehicle. The steam boiler drove a generator that produced electricity to power an electric motor to propel the vehicle. Nothing today, it seems, is new.

Needless to say, the steam-driven and petrol-driven omnibus, all hybrid in configuration, was 'painted' in poor light by those who owned and ran horse-drawn hackney cabs and public service alternatives. The powered omnibus was presented, through propaganda, as dangerous: liable to explode and become unstable, even to travel too fast. There was a point when society considered that if the human body was propelled at more than 12 miles per hour, physical damage would be done to the human frame.

"Do not confuse me with the facts; it is what I perceive that is the truth." Mountains out of Molehills (TPO'H) 2003

Nothing changes, does it?

Vested interest and hypothesis permeate and restrict progress. Those who have a position in, and guide the marketplace, inevitably resist change - no matter how good that change might be for the planet or the public at large. Opinions varied then and still hamper real progress today. Capital expenditure and its amortised period to a financial return affects judgement.

It fell to a man named Walter Hancock to take up the challenge. He followed George Shillibeer and his pioneering work with the horse-drawn omnibus by way of a vehicle named 'The Enterprise', which was steam-powered and electrically propelled. The service operated from London Wharfe to Paddington. The problem seems to have been that steam carriages were too far ahead of their time. The government did not help by imposing taxes and tolls on steam vehicles, and, as already mentioned, vested interest made sure that life was difficult for those trying to introduce the new technology. Could that happen in the 21stcentury?

Another big problem was that carriageways were not entirely suitable for those inherently heavier vehicles, which those with the boiler, plus twenty passengers, would certainly be, compared to the lightweight Hackney cab and similar horse-drawn facilities. The rumours were not all fiction regarding the safety of alternatively powered vehicles; for example, in 1830, a steam-powered vehicle partly demolished a house when it ran off the road. Accidents did happen then and do now, vis a vis the onboard battery fires on electric battery driven buses. Fear of infrequent happenings often distort common sense appraisal.

Horse-drawn trams were in popular service from 1860, and electric trams entered with their overhead power line system as a natural adaptation in 1885. Mechanical advancement had changed the pattern of horse-drawn bus services. In 1900 there were still 3,676 horse buses in London.(3)

***4.1** The basic horse-drawn bus.*

The first horseless carriages really made their mark, probably in London, around 1926 with petrol engines and 'specials.' It was the advent of the internal combustion engine that changed everything. The first motor buses using the new petrol engine generator mechanism came into service in the early part of the 20th century initially as hybrids with electric drive propelling the vehicle using the electricity generated by a petrol engine.

It was much later that a direct drive from the internal combustion engine was introduced following the successful development of a gearbox. That innovation meant that the electric motor was displaced from the vehicle as unnecessary weight. It followed, therefore, that from 1909 onwards, when the main bus companies actually made profits from these vehicles, that the end of the horse-drawn era was assured. By 1911, only 30 horse-drawn buses remained in service in London.

"The motorcar will never usurp the place of the horse," The Economist 1911

But it did. And it did it 'big time'.

Chapter 5.

The Appearance of Electricity in Transport

Passenger-carrying 'carriages' on rails were certainly the major innovative step to modern energy development in the Victorian era. When Queen Victoria came to the throne in 1837, Buckingham Palace was lit by candles. It was still lit by candles towards the end of her reign. The environmental impact of candles on the buildings they lit and the lungs of the individuals who lived in the properties that depended upon them is well-known for its detrimental character. Respiratory disorders abounded. Even so, it still took many years for gas lighting and then eventually electric lighting to become the British nation's norm – even in 'Buck House.'

When gas lamps became the desired 'norm,' gas mains were installed along the many miles of streets entering each house and factory along the way. However, much was new build; therefore, incorporation of that development was easy and relatively cheaper than the modernisation of existing stock. It was there to provide not only domestic heating and cooking facility but was also utilised on the streets themselves to illuminate the dark passages as well as the broad thoroughfares.

The change was not taken lightly, and the cost of that change was not cheap; however, it coincided with domestic running water and sanitation. Water mains and sewers were major infrastructure innovations at the same time. Digging up the streets became fashionable. The result was improved streets. Pavements were created for pedestrians. The road surface was improved. All as a by-product of innovation and modest, if temporary, inconvenience. Who remembers Diamond Cable being dug into our pavements? The introduction of broadband.

Was the environmental benefit of the change really understood or accepted by the people who were affected by it? Society still uses that infrastructure. Gas lines are being renewed and upgraded in 2021. Ceramic and steel piping is being replaced by continuous plastic material to facilitate the use of natural gas in the 21st century. The original infrastructure cost has long since been recovered, and the benefits to society, as a whole, are not often appreciated. Familiarity does breed contempt. New infrastructure expenditure anticipates the continued requirement for the gas, water and sewerage to be transported. The capital cost to society is amortised appropriately over a significant time scale for recovery.

Certainly, the lights burnt brighter and less of the sources of light energy were needed to produce the superior illumination achieved by burning gas. However, where was the gas produced? The gasworks.

Gasworks, like the steelworks, depended upon coal being mined and brought from those mines by steam locomotive drawn trains, on their steel tracks, to the gasworks. The coal was effectively converted, by a 'burning' process, into gas for distribution to the public through the underground pipe network that was laid as part of a large town or city's infrastructure change. While those pipes are being replaced, the work is intrusive. The rest of the time, the system is taken for granted.

Has there been a community or environmental improvement resulting from the installation of gas, now natural gas, in those communities? What are the pros and cons for the people who live with the facility and use its powerful resource?

A by-product of coal gas production was coke.

Coke is a 'smokeless' fuel that has played a significant part in reducing the sulphur laden smoke that came from burning coal in furnaces and on domestic fires. This smoke was responsible for 'smog,' that green fog that was all too prevalent in many urban conurbations in the late 1950s. Coke, however, was not universally adaptable. Sulphur is a toxic effluent gas that contaminated the atmosphere from the chimneys where coal fires were the norm and in so doing 'ruined' the atmosphere whilst providing efficient fuel to cook with.

'Smog' was a contentious by-product of that sulphur effluent. Natural gas is an answer to that detrimental aspect of the old gas production method. The question has been asked, could that natural gas also be burnt efficiently in the internal combustion or compression engine propelling public transport vehicles? The answer is in the affirmative.

Coal was also used to fire the boilers in most of the electricity generating plants in the UK up to the 1990s. This created the 20th-century form of power that was to enter most homes, in most streets, comprising most major conurbations and replace gas in many situations, including cooking.

Electricity was, and remains, the new 'clean' power source that was to become the central means of providing light and a motivating force for almost everything in society over time. There was, however, a lingering environmental hazard concerning electricity's production mechanism. Was that hazard so environmentally unacceptable when compared to the steam engine, the open coal fire, gaslighting, candlepower, and all else that had gone before? It is an interesting question, and the answer is still, apparently, being assessed.

The effect of electricity, in transportation terms, was to become quite notable with the advent of the tram as an electrically powered inner city means of transporting people in the early part of the 20th century. That change of emphasis was replicated on the railways, where electrification took away the likes of the underground steam locomotive system operating out of Baker Street in London. The conversion of the line to a high-speed electrically driven passenger train service would effectively transform the means of travel, not only in London, but in New York and Newcastle, and other major cities around the world. Steam and diesel locos were still in operation on the London Underground in the 1960s. The author travelled on them. The transformation of a major national network takes time. Time, now, it transpires, is something we must not take for granted.

Overland train systems were converted to electrical power as early as 1890 on the London underground system. That conversion continues to this day with a variety of configurations from the third rail Southern railway system through to the underground 'fourth rail' and then north of London and south-west where an overhead wire system powers some two-thirds of the rail network's electrified system.

That conversion work to upgrade is still the subject of controversy and remains ongoing in its introduction to diesel-electric train operating routes. Once again, the hybrid mode of propulsion features in the diesel-electric locomotives that replaced the steam engine. The power produced is steady, strong and sustained in use, eminently suitable for heavy haulage.

It is noteworthy that in 2007, the government white paper *Delivering a Sustainable Railway* favoured biodiesel-powered trains ruling out any large-scale electrification infrastructure program for at least five years. A belated change of emphasis took place, resulting in Great Western and The Midland network being electrified with overhead cable delivery by way of a single strand catenary.

Interestingly the overhead line rather than third rail was chosen as the order of the day as it is more economical and permits faster train speeds. Electric motor driven trains have power advantages regarding smooth acceleration and refined braking. These are attributes that suit them to the haulage of heavy loads. Passengers acknowledge a cleaner air environment and smoother journey; however, the advent of diesel over steam and electric over diesel still creates intense discussion as to which is favourite overall.

Double-deck tram units became commonplace in many cities in Great Britain and Ireland. They operated above ground in those cities with some interesting variations, such as along the London Embankment and then turning north to proceed underground up to Holborn and into the centre of London approaching the city itself. In the USA, Boston, Massachusetts, also had a double-deck subterranean tram system into Boston's city centre.

The trams were fed electricity from a single overhead cable grounded through the steel rails that the two-storey trams ran on. That system proved to be a clean, relatively cheap and certainly effective mechanism to move large volumes of people from one place to another with a small footprint vehicle. The tram's main 'Achilles heels' were, and remain, its lack of flexibility and inherent noise, along with the rails in the road surface. Tram systems are costly to install and considered hazardous to road users.

Maintenance costs of tramways became an issue over time as rails required replacing, and, of course, the tramlines created a dangerous route hazard, particularly for cyclists and motorcyclists alike. Even cars have been known to 'tramline,' i.e., get caught in the influence of the rail line profile. In the mid-1920s, tram chassis were adapted to ordinary carriageway configuration, fitted with pneumatic tyred undercarriage (apologies for the aeronautical term) and a twin pole overhead power pick-up to become trolleybuses. Modern trolleybuses are structured around bus chassis designed for the purpose to give comfort, stability and silent running.

5.1 *Converted tram to trolleybus.*

Chapter 6.

The Tram with Pneumatic Tyres and no Rails to Run On

Experiments were carried out in the 1920s to give flexibility to public transport whilst maintaining the environmental kudos of the electricity-powered tram system. There were no exhaust emissions from trams to concern the populace nor the authorities regarding air pollution or disfigurement of the buildings in the environment that they operated in. The search was for some mechanism by which the rails could be removed to increase road safety and reduce pedestrian/cyclist peril and yet maintain the cleanliness of the tram's mode of public transport in urban centres.

The bicycle was a development in the Victorian era in much the same way as the other forms of transport that were being developed. Its pertinence here is twofold. The bicycle became the personally preferred method of transportation to carry people to a place of employment as it was acknowledged as a relatively cheap creation for transport independence. Travel to senior schools by bicycle also came into that orbit of thinking and application. Maintenance and running costs were relatively cheap for bicycles. However, the bike did have its problems.

Accidents frequently occurred on busy streets during ‘rush hour’ movements when literally hordes of cyclists vied for position on the routes to or from their employment. ‘Speeding’ and careless riding were not the sole prerogative of the motorist. Even today, cyclists can and do become extremely competitive, if not combative. Tramlines were a perpetual hazard in major conurbations where cyclists failed to recognise the dangers that the rail groove structure in the carriageway presented if the cyclist was distracted. Pavement kerb edges also presented problems along with breaks in the road surface.

The problem with tramlines running in inner-city and large town environments in the 1920s was that most people moved from place to place either on foot or on their bicycles. Tramlining became a common cause of major accidents, resulting in broken limbs for cyclists and pedestrians alike. The twin rail profile, unlike the single rail profile employed with trains, created a groove into which the cycle wheel would drop with no effective means of escape other than for the rider to dismount and lift the bike out. Fatal falls were not uncommon.

Any cyclist will testify to the difficulty in performing that ‘wheel extraction’ feat of gymnastics. Dismounting from a cycle in the circumstances described was seldom, if ever, accomplished gracefully - with one wheel in the track, probably the front wheel, the steering wheel, and the other outside of the track. Other practical hazards militated against the wider use of trams; the heels of shoes could also be caught in the track, resulting in leg and ankle injuries. The infrastructure cost was also a major governing factor when deciding on the installation of a tram system. Tracks are inherently expensive to install and maintain.

In the early days of electricity production, the use of direct current (DC) power supplies dominated the power provision required to propel a vehicle. The tendency was to use DC electricity in the early years of development, not only for the vehicles but for the electrical supply in general use. There were specialist shops at which to get the accumulators, a type of early battery, charged in order to power a domestic radio certainly in the mid-1950s.

Thought was given to a positive and negative wire placed above the road that the vehicle would be powered by. This would have been an alternative configuration to the single overhead wire and single collecting pole (or boom) mechanism of the tram system and the system used on the railway network, the pantograph.

As early as 1882, experiments were conducted where a twin pole power collection system was used to provide the positive and negative 'terminals' for the electric drive motor. This eliminated the requirement for the vehicle to run on rails in order to provide the completed electrical circuit required. If that twin-boom configuration could be achieved, then a viable alternative to the tram would be facilitated. And it was. That achievement led to the conversion and subsequent replacement of many tram vehicles and thereby tram-based services.

The initial, practical use of the twin-boom/overhead power cable system gave its name to the vehicle employing that twin feed cable system. By having a small 'trolley' (wheeled trailer) running on top of the twin cables and collecting the electrical charge, which then conducted electricity to the vehicle below, the electrical circuit was completed. That electricity was fed through to the electric motor, which powered the carriage style vehicle forward, dragging the trolley along the overhead cable. The negative line completed the electrical circuit.

***6.1** The original trolleybus – note the overhead 'trolley' on the wire.*

That system proved adequate for a vehicle travelling on a system with a starting point involving a single destination. However, problems arose concerning the principles of the mechanism. The 'trolley' would have difficulties negotiating junctions or, more importantly, running smoothly over any device designed to keep the cables at a uniform distance side by side and in contact with the 'trolley' to vehicle electrical contacts. The mechanism worked over relatively short distances and perfectly in fine weather.

In order to facilitate the use of junctions and other 'line hazards,' the twin cable system would be used but developed in such a way that the connection to the overhead electrical supply cables was facilitated by poles that would be positioned to have the electrical connecting pieces pushing up from underneath the wire rather than the 'trolley' feeding current from above the overhead power cables. By resolving that requirement, the modern 'trolley' bus was born, without the need for the 'trolley' that gave it its name. The 1920s and 1930s heralded a major change in the transport system employed within many major population centres, and the trolleybus was fundamental to those changes. A new era was born.

Just as the railways had been created with an infrastructure that provided the facility for movement between towns and cities on a large scale, the trolleybus infrastructure was put in place along set routes with limited but quite useable flexibility, without the need for rails embedded in the thoroughfare. The roads became safer, and the electrically powered vehicles themselves could pull into the kerb to allow passengers to board safely whilst the trolleybus remained connected.

Due to the mobility inherent in the power connecting poles' base units and connecting heads, the trolleybus could also 'pull out' to the centre of the road to pass vehicles on the nearside or to avoid road works; something that a tram could not achieve.

Petrol engines ran on leaded petrol and therefore carried an environmental stigma because of the 'health hazard' derived from the lead in the petrol being emitted in the exhaust gases affecting health. As with so many negative manifestations, it was only through usage that the detriment was discovered. Petrol power was temporarily out of favour.

Mechanical junctions in the overhead twin line configuration were developed in order to provide the required flexibility allowing varying routes utilising the same overhead lines whilst allowing for competing routes to diverge. This meant that a fully diverse system could have vehicles setting off from a common point with a wide variety of routes available as the system moved away from the centre. Similarly, routes could conjoin and run common routing and then diverge to separate destinations. The permutations were fascinating and practical.

The effect was electric. But then; the government nationalised the electricity generating enterprises. The cheap electrical power that propelled the trolleybuses became an expensive element in the then capital cost equations as the catenary life demanded renewal and the bus fleet required updating. The timing could not have been more significant.

For over forty plus years, the trolley bus provided a major source of transport in most of the major cities and towns throughout the United Kingdom. The end of that era was brought about by the advent of the diesel-engined public service vehicles and the circumstantial costs of electricity and infrastructure upgrading.

The diesel engine's introduction into bus manufacturing would propel those large vehicles more economically than the petrol engine, deliver more flexibility in use than the trolleybus and have little infrastructure requirement to consider. Being less specialised in operational format, the diesel bus was cheaper to produce and would provide for deviation from a given route, should the need arise. The fleet was deemed to be more flexible, and it was.

The coupling of the operational advantages with the production line savings of a comparable capacity vehicle by way of its relatively lower cost of manufacture and infrastructure at a time when the trolley bus fleet and its infrastructure required renewing meant that the arguments for change became an economic certainty. The diesel engine was 'a cleaner', less thirsty and more powerful unit compared to the petrol engine. The environmental advantage of diesel power over petrol-engined vehicles with the added economy in miles per gallon of fuel meant that the die was cast for a radical change of emphasis.

It is perhaps pertinent to note that the trolleybus was much cheaper to run per operating mile than either petrol- or diesel-powered vehicles by a considerable margin by comparison. Analysis shows that the breakeven marker came about at the 25-year point. (1) Although the initial capital cost of the diesel vehicle and its infrastructure was cheaper, the actual amortised operating cost over an extended vehicle lifetime was almost identical. A further consideration in the 1960s decision was the nationalisation of the diesel bus manufacturing facilities in the UK.

The introduction of the diesel engine public service vehicle gave way to the removal of the overhead power lines from towns and cities along with the need to maintain them. Diesel engines were declared 'clean' compared to the petrol engines which in those days used leaded petrol fuel of varying grades. The lead-free petrol that is now used in petrol engines had yet to be developed.

Neither the petrol nor diesel engine vehicles could claim to be as environmentally clean as their electric counterpart, but in the 1960s, it was perception that carried the debate. History was, it transpired, going to prove that point with a vengeance.

Chapter 7.

The Presumption Regarding the 'Clean Air' Compression Engine Was to Last over Forty Years

We are now experiencing a complete reversal of thinking that takes technology back more than 100 years.

The first petrol engine buses were apparently created by General Electric, an electricity generating company. The involvement of the petrol engine was to provide the required power to an onboard electricity generator that fed the current produced to an electric motor that drove the vehicle. The weight disadvantage of the two entities was outweighed by the smooth running of the vehicles fitted with them. The absence of the need for gears made the operation of the vehicle smooth and powerful.

In those early times, for the internal combustion engine, a gearing mechanism was essential to cope with optimising the transmission of power to the wheels. The intermediary 'crash' gearbox was a rudimentary and acceptable answer that had not been invented, let alone perfected in the 19th century. It was to be after the First World War that gears were developed sufficiently to make a direct drive facility possible. It was after the Second

World War when the preselect and synchronised gearbox became a recognised feature of gear change mechanics that would give the internal combustion and compression engines their head.

A lot of the early petrol-engined lorries were flatbed trucks, which, when adapted for public transport, gave way to the adoption of the word charabanc. The lorry chassis had a flat platform behind the cab onto which goods could be loaded and transported. A detachable platform of forward-facing bench seats could easily be created and fitted onto the flat back of the truck with a step style ladder at the rear to allow passengers, perhaps 10 to 12 in number, to mount the platform and be seated for carriage 'to town.' An alternative mode of transport was created.

The compartment would have had simple bench seats and eventually became enclosed, creating a permanent adaptation for specialised vehicles.

Apparently, one of the prime operators of that format of public transport facility was the coal delivery man. They would transform their coal lorries (flat trucks) by mounting the cabin facility at the conclusion of their coal delivery work. They would then operate the 'omnibus' service using that same vehicle by putting the charabanc on the back with a ladder to enable access to the bench seating. All of the early buses comprised coachwork mounted on commercial vehicle chassis, which developed over time.

A typical petrol bus in an industrial town would be powered by, say, a four-cylinder Gardner engine propelling a Tilling Stevens single-decker bus. That vehicle type was replaced after 1947 and onward by the more successful and stylish 28 hp Bedford coaches that boasted a 29-seater 'Duple Vista' body which the author recalls readily from his 1950's childhood. The Duple coachwork had that sleek, sloping back. Those vehicles were originally developed during the Second World War as troop moving vehicles. It would have been 1952 when such coaches were used as school buses for infant-aged children to get to school.

The larger 33-seater single-deck buses were again Bedford chassis fitted with Duple bodies, and they would achieve about 10 to 12 miles per gallon of petrol. This was along the lines of the SB vehicle, which came into service after 1951. That coach version was notable due to the full-width flat front (no bonnet) and engine in the cab. The engine was next to the driver, and they were operated by the likes of Royal Blue.

These were a regular sight on the A30 London to Southampton daily service through Bedfont, Middlesex. Royal Blue was a very large company that was nationalised in the late 40s under the Clement Attlee government programme. It was the same time that the steel industry, the rail industry and the large haulage operators were nationalised. Smaller firms could operate coaches, and many small haulage companies were similarly allowed to operate independently.

The larger double-decker bus fleets operated 'Bristol' buses. This was a chassis fitted with beautiful Eastern

Coach Works bodies. They were diesel engine driven. Diesel engines started coming in towards the end of the war as vehicles became larger. AEC diesel engines set the trend and were in a prime position regarding the introduction of the mass-produced British diesel engine market. Most of the nation's double-decker buses were diesel engine driven at the outset of the 1950s.

Vehicles in the public service sector came in two distinct types, heavyweight and lightweight. Heavyweights invariably had diesel engines, whereas the Ford and Commer commercials, based upon a small lorry chassis, had six-cylinder side-valve petrol engines, much the same as the larger saloon cars from the same or connected manufacturer. The Austin Princess engine was fitted in Austin built lightweight coaches. Vauxhall also developed a lightweight vehicle with a six-cylinder petrol engine as used in their own large Vauxhall cars. There was a diversity of manufacturers and competition in the market.

The bigger-bodied Bedford had 33 seats and would achieve similar miles per gallon from using the diesel engine as the petrol engine. However, diesel engines were powering the heavy Leyland buses, which were a lot more robust than their petrol counterparts. The larger buses made operational savings in maintenance costs which proved to be substantially less than the maintenance costs of the lighter-bodied vehicles. Again, the principle of amortising costs over the expected lifetime of the vehicle proved economically astute.

The suspension springs on the smaller vehicles had to be renewed every year, whereas the larger vehicles were only required to change the suspension units every three or four years. The same applied to steering joints and so on. Those heavyweight public service vehicles were million-mile vehicles. Once again, the amortised capital costs coupled with operational costs proved to be worthy of analysis.

When Leyland got into financial difficulties, the firm was nationalised. The Leyland National bus was developed by British Leyland with the National Bus Company (NBC) and entered service with the NBC subsidiaries on a regional basis. The Leyland National bus replaced many of the Bristol-made buses and the trolleybus fleet over a very short time span. However, the Leyland National buses proved to be a poor substitute for many reasons.

The Leyland chassis was of poor quality. The operating costs grew exponentially due to the Leyland's high fuel consumption. Even though the Leyland National bus was diesel engine driven, as a vehicle, it achieved a mere five or six miles to the gallon. The buses were too big for the country roads they were employed on. The sales and production of the vehicles were controlled by the nationalised company operating the vehicles. Therefore, that company purchased the similarly nationalised bus fleet from Leyland. The Leyland National 1972 coach and Leyland Leopard and Tiger vehicles were widely used and just as widely criticised.

The Bristol SC was a lightweight aluminium chassis providing a 33-seater configuration with the Gardener diesel engine. It was ideal for town service given its economical operating costs and fuel consumption of 15 to 16 miles per gallon. The Leyland National buses were half as big again and therefore unsuitable for many suburban and rural routes with the added impediment of more than twice the rate of fuel consumption. Nationalised bus operations were a feature of public service vehicle provision between 1969 and 1988 that was simply lacking in commercial acceptability.

The nationalised entity owned the subsidiary operating companies that provided the public bus services. It did not run buses itself. London Transport and the Tilling company were nationalised in 1948. THC (Transport Holding Company) was the holding company. British Electric Transport sold their 25 provincial companies comprising 11,300 vehicles to the state in 1967, which then meant that 90% of public transport became the National Bus Company with effect from 1st January 1969. By that time, the major bus manufacturers were all owned by the state; Bristol Bus company and Eastern coachworks (75%) plus British Leyland (25%).

The situation was eventually reversed, and those industries denationalised. It was Margaret Thatcher who deregulated the industry, and independent minds took control under the 1986 deregulation protocol. Competition returned to the bus manufacturer market. However where

British Leyland had once led the field in diesel engines, those that produced the engines under licence, Volvo and DAF, developed and improved the Leyland vehicle engines to a higher specification and took the market to another innovative level. Leyland was left behind.

Once more, the focus of this discourse is obliged to return to history. After Shillibeer's launch in the omnibus public carriage market, smaller buses came into use carrying just 12 passengers in a small box-shaped coach pulled by two horses. An improvement on that original bus format with a raised roof entered service on 9th May 1847. This had a large doorway which made it easier for people to board the bus.

As competition grew, operators introduced vehicles with lighter frames like the 'knife board' horse-drawn 'bus.' It acquired this name as the seating arrangement on the top deck was similar in profile to the wooden block used in kitchens to hold kitchen knives along the grooves in the board. The passengers would sit on long benches configured back-to-back, which resembled the knife boards.

The conductor would be perched on his platform at the back of the vehicle. They operated in service between Putney Bridge and Piccadilly London from 1859 until well into the 1870s when they were replaced by the 'garden seat' type of vehicle from the 1880s onwards.

7.1 The hybrid petrol-engined electrically propelled bus of yesteryear. (LAO'H)

Entrepreneurs are often the people that suffer the most from progress, and no one suffered more than George Shillibeer, who was declared bankrupt and unable to pay his debts. He went to debtors' prison. He spent the last years of his life as an undertaker, having adapted his bus for use as a hearse. He lived to the ripe old age of 69, nonetheless. British Leyland, the nationalised entity, suffered a similar humiliating end.

Just as today, it was the congested narrow streets that created problems for public service vehicle operators. However, one inventor came up with a novel idea to surmount that impediment. In 1839, Mr W. B. Adams constructed a vehicle comprising two parts joined in the middle by a flexible leather passage to enable the vehicle to turn more easily on the narrow streets. This was the original 'bendy bus' - the forerunner of today's modern manifestation. Nothing, it seems, is new. What is perhaps 'new' is the requirement for a huge turning circle by the 'Powertrain' multi-articulated mass transit vehicles operating in Europe - quite the opposite of Mr Adams' aspiration for his vehicle.

History tells us that in the United Kingdom in 1931, 4,754 buses had carried nearly 700,000,000 passengers over 185 million miles. Each bus would employ a driver and a conductor, the latter collecting the fares, and most of this activity occurred in London. The transportation of fare-paying passengers became a major employer with over 32,000 staff across the nation.

The development of the bus has allowed the rank and file of society to exercise mobility on a scale never achieved before. In the 1950s, public bus services could carry a family from rural Middlesex to a day out at Whipsnade Zoo or to a once a week performance at the Chiswick Empire, or Shepherd's Bush Empire, or even along to Hammersmith where a change of bus enabled the journey to Hackney Empire to see Joe Loss and his band. All of that public movement had been commonplace since the 1930s in cities across the United Kingdom. This was pre-television and the advent of common ownership of a car. The public transport system up until the 1960s in almost every major city in the UK comprised a wholesome fleet of electric vehicles - trolleybuses - that gave clean air, a smooth ride, and trouble-free, low maintenance service. They were in motion charged electric vehicles by another name. That was all to change.

The bus routes increased and improved around most major cities during that time. Coach services were also introduced, with far-flung destinations being reached in hours rather than days. It was the time before the motor car became popular (and affordable) with the general public.

Public transport was widely used by a huge percentage of the public up to that point. Of course, that was to change as motor cars became more widely owned and independence in travel became more the norm. However, the novelty of driving and the pleasure that it once gave was to be adversely affected by the volume of journeys undertaken. Today we have the spectre of electric vehicles being an expensive alternative to fossil fuel cars. Could public transport become more 'the norm' once again?

Chapter 8.

History Is Repeating Itself

Having considered history up to the advent of post-war Britain, the question remains: is the past and its progression relevant to the future?

A far more pertinent question concerns those that make decisions about public transport in the United Kingdom and their understanding of the historical usage of the public transport systems as they developed over time. History is pertinent, indeed essential to the decision-making process in analysing what is best for the future. Not in a Luddite sense, but in a common sense.

Most of those making judgements will certainly have never travelled on a stagecoach style carriage. However, the Duke of Edinburgh certainly did and could probably testify to the 'thrill of the chase,' as well as the lack of 'comfort' that might be afforded to the travelling public in such vehicles. Of course, some of us will have driven motor vehicles 'off-road' on farm tracks, perhaps, but how many of those making infrastructure decisions will have made a long journey on those rough tracks? Few, if any, would constitute a good guess at the answer.

The point being made here is that in order to make an accurate 'assessment' and subsequent 'judgement,' one must perhaps experience 'the topic' that one is endeavouring to analyse. In this instance, the author claims that position.

Few of the generals leading the WWII South-East Asia campaign against the Japanese realised the difficulties of the terrain; the topographic impediments encountered by their troops tasked with missions dictated by those military commanders. It was only when those in command experienced, saw, felt and understood the environment for themselves, and thereby the errors of their tactical assumptions, that realisation dawned. Much unnecessary detriment had already been experienced from map-based conjectures rather than informed assertions based upon situational fact.

So it is with transport theory.

Several years ago, an experiment was conducted by a businessman who decided to leave his car in the railway station car park and use public transport over a 48-hour period to visit six of his clients in the south of England. The journey started in the East Midlands. It was an experiment designed to establish first-hand the efficiencies or otherwise of the public transport system and the cost of using it. It transpired to be an extremely informative exercise.

The initial train journey to King's Cross was a standard procedure; therefore, the experiment really started by way of the short walk across to St Pancras Railway Station in order to commence travel on an express train to Ashford in

Kent. Ashford is where the experiment truly started, and its pertinence to this discourse becomes relevant.

The 'country' single-deck bus (a coach by DfT definition), which was to take the businessman out of Kent and into West Sussex, was not timetabled to meet the train. A 30-minute wait was all that it took to herald the arrival of a single-deck diesel-powered 'bus' with its final destination clearly defined as 'Hastings' in order to get our intrepid traveller to his first call at Battery Hill, Fairlight, near Hastings. As can be assumed, the meeting with an old friend was cordial, and the time of the next bus into Hastings carefully noted for the duration of the meeting to be established and the journey to continue.

Hastings to Brighton completed the next section of the journey by double-decker to facilitate the second visit on what was to prove a very busy excursion involving a further scheduled transfer to the Hove bus, which was executed like clockwork. The timetables really did work well. Late lunch in Hove sent our intrepid traveller on the last part of the first day's journey by bus to Steyning. Then there was an overnight stop that included a delightful dinner with his third friend and colleague, accompanied by delightful conversation bringing all the relevant personal news up to date.

The second day's journey started mid-morning by train and ended in Crowborough with a short bus journey requiring a short walk from the main road to the client's house. Just as before, plenty of time was allowed before the next bus trip was due to be engaged and, with lunch included, the brisk walk back to the bus stop was uneventful. The bus duly arrived on time, and with a mild hiccup which involved being collected by the subject of the fifth appointment, that Tunbridge Wells appointment proved to be relatively incident-free. A late afternoon train back into London ensured passage on the 7:06 pm train back up to the East Midlands, concluding a very successful experiment with time allowed for dinner at 5:30 in the 'Betchamin Arms,' St Pancras station, to achieve a sixth appointment prior to boarding.

Not only did the experiment prove that one could travel around the United Kingdom quite cheaply, but more importantly, it proved the progress that has been achieved in public transport provision in the last 150 years. The experiential and entertainment value of the journey by way of a contrast to that same journey had it been conducted with our businessman driving his car is also worthy of consideration. Even as a passenger in a car, the traveller would have seen and experienced a lot less of the passing countryside and his fellow travellers than was accomplished by his journey on public transport. That surely made the experiment worthwhile and worthy of note.

The trains were fast, punctual, and covered long distances comfortably, quietly, and in good time. The passengers were unaware of the sound of the diesel engines of the '125' that they travelled in on the outward and return journey involving King's Cross. The express from St Pancras to Ashford in Kent, the Eurostar express, was electrically driven and even quieter to the outside world than the '125,' even though both ran on continuous rails.

Both forms of transport provided an excellent view of the countryside from the raised position of the railway network along most of the journey. That certainly changed on the bus network, particularly so on the single-deck transport from Ashford. Beautiful countryside was obscured by hedgerows which the top deck of a double-decker bus would have provided wherever the traveller sat.

It appears to be a fact that a significant number of people use the bus network just to get from A to B. However surely there are also significant numbers of pleasure-seekers in that the journey is, or at least has some facility, to make it more enjoyable. The single-deck rural buses were extremely functional compared to the top deck of the double-decker buses used on most of the routes experienced, which provided transport for much of the journey. The single-decker was functional but mundane. The top deck became educational and enjoyable because of what could be seen as the journey progressed.

It can surely be appreciated that some of the people on those original hard wheeled carriages moving over cobbled or poorly paved streets felt excitement brought on by the difference in the experience provided by their transport whether or not it was pulled by a horse or engine driven.

Similarly, those on the open top deck of those converted platform lorries by way of 'make do' local buses would have experienced an element of excitement seeing the world from a higher vantage point than was the norm, coupled with travelling 'at speed.' Only those on tall horses would have experienced anything similar to it.

Surely the point of travel is that, apart from being economically viable for the consumer/passenger to get from the point of boarding to a given destination, it should be as exciting as it can be. In our day and age, that perhaps incorporates well-sprung, pneumatic tyred, quiet and climate-controlled vehicles that are a pleasurable way to get to our destination.

Sitting in the downstairs salon of a Leyland double-decker bus in those rear seats which face each other adjacent to the rear engine housing and the heat and clatter that accompanies the journey can pass for comfort. However, it can surely be improved upon, first by removing the noise of the engine and also by removing that unbearable heat on a crowded bus, particularly in the middle of an otherwise hot day.

That simple transformation can be achieved using electric motive power, which brings the discourse to the 1990s and the reintroduction of the hybrid method of motive power use in public service vehicles, almost exactly 100 years after *'The General Electric Company'* used it in 1897.

There is no doubt that the theory behind the hybrid is faultless in terms of the smoothness of operation of the vehicle. However, consider if the electric motor was powered from a battery, as some are, charged from an in-motion catenary system in the urban and city elements of the journey. The quietness achieved within the vehicle as a whole, as well as to the outside world as it passes, would be a significant experiential upgrade; married, of course, to the exceptional environmental benefits that should follow on.

But do they, the environmental benefits, "follow on?"

The problem with the hybrid is twofold. The batteries do not have the capacity to propel the vehicles without the intervention of the fossil fuel engine backup to achieve an acceptable range. The vehicle battery life is restricted to 3 or 4 years due to the operational mechanics requiring constant recharging of a public service vehicle with its frequent stops for passenger and traffic interventions. Average speeds of 4-7 mph are not uncommon. Hills are also an impediment.

There is also a common disadvantage that modern batteries Lithium-Ion - also NMC, NCA, LFP, LTO and the upcoming "solid state" batteries in that they are not recyclable. There are speculative reservations on that topic. Schools of thought imply that there is a further application for the replaced batteries. However, is that 'further application' cost-effective environmentally or fiscally in the longer term? Is the notion perception, hope or fact? There is also the overriding question pertaining to the availability of the relevant mineral component parts required in modern battery construction. These are finite resources with an environmental cost for mining and extraction which must have relevance in light of increasing demand over time.

As has been shown, history moved from the hybrid to overhead power supplied tram systems which were more reliable but dependent upon a noisy steel rail-based technology and the inflexibility that accompanied it. Transport on high usage journeys progressed to the pneumatic tyred overhead power supplied in motion electric bus and that surely has to be a consideration in the 2020s.

What we are experiencing in the 21st century is a complete reversal of the power plant appraisal where public service vehicles in particular and heavy goods vehicles in general are concerned.

So why do we find ourselves in 2021 (or even 1998, for that matter) with a vehicle propelled by an electric motor using electricity generated by a conventional fossil fuel engine? Would a better transition for public service vehicles be to rely upon full electrically powered vehicles with the ability to range 'off wire' on battery power? Or possibly even extend the operational capacity of the vehicle beyond battery power into non-urban areas using diesel power as they do now? The Euro VI diesel engine is much improved in so many ways, and perhaps the "needs must" mantra cannot be ignored, at least until we have manufactured a satisfactory replacement.

The hybrid public service vehicle has an impediment in that the duration of pure battery-powered travel is relatively short. The government and activist endeavours seem to be to eliminate the exhaust gases from the fossil fuel engine, whether it is diesel or petrol powered, coupled with the increasing use of environmentally friendly centrally produced electrical power. This would seem to fit that same requirement as for vehicles that have a prescribed public function to operate over a known route on a consistent basis.

The reintroduction of overhead power lines (the catenary) would be a relatively cheap infrastructure change to accommodate retrofitted hybrid vehicles, thus converting them to an electric bus with onboard battery charging for suburban route extensions, decoupling, and recoupling at will. The system would safeguard other road users, such as cyclists within urban city areas, and it would be cheap to operate and maintain.

Historical data informs us of these facts, as do the learned papers referred to. Historically the In-Motion Charging and electrical supply system trolleybus system is a proven facility that has merely been ignored in the United Kingdom, largely based on the false premise that diesel is clean and as a result of a nationalisation program that bypassed it.

The most economical mechanism to facilitate the elimination of toxic particles in densely populated urban locations must surely be the reintroduction of the in-motion trolleybus principle into the United Kingdom public service network. History indicates that is a rational and natural conclusion as it demonstrates its effectiveness whilst the quantum leap to hydrogen power is developed and goes into production. This can so easily and cheaply be achieved by retrofitting the current hybrid buses to draw electrical power from twin overhead cables, the catenary.

That catenary-based facility would enable the design and manufacture of UK created double- and single-deck trolleybuses to take place - if that was deemed the intermediate solution of choice. It may well extend to a three to five-year or longer process to allow for any full transition into hydrogen or other technology in 2040 and beyond. The immediate change to incorporate overhead power provides time whilst addressing the urgent environmental dilemma that cities face now.

The 'north of Watford' cities can benefit most from the retrofitting of their current hybrid bus fleet. The vehicle fleet's operational life extension alone will come near to paying for the modification.

By creating an urban overhead power infrastructure facility - a catenary - to facilitate in motion electrical charging, local authorities would promote a smooth and swift transition to electrically powered vehicles in densely populated areas. The catenary facility would obviate the requirement for battery-powered vehicles to interrupt their journeys to recharge the onboard batteries, thus improving the journey time and ultimate passenger experience. A meaningful transition could be achieved within 2.5 to 4 years for the 1528 vehicles involved outside of London.(7) All of that work can be developed and manufactured in the UK or through its trading partners, creating jobs and innovative development. Surely that is a prize worth working towards.

Chapter 9.

An Embraceable Leap Forward

The advantage of the retrofit adaptation to the current hybrid fleet is the improvement in versatility within the public service vehicle (PSV) fleet. The ability to connect and disconnect 'at will' to overhead power lines improves electrical bus operational performance, enhances battery longevity, and thereby saves capital expenditure amounting to hundreds of thousands of pounds over relatively short time periods. The vehicle efficiency when considering traversing undulating terrain speaks volumes for increased efficiency and certainty.

The proposed adaptation creates an operating mechanism that can start off connected at one point to the overhead power source that provides the driving power whilst concurrently charging the onboard batteries as applicable to 'all electric' battery-powered vehicles. A diesel hybrid bus can then disconnect from the catenary and proceed in a rural environment using the diesel engine as appropriate.

Once the route enters an urban environment, the vehicle can operate by a combination of catenary sourced overhead electricity or onboard battery into the centre of the town or city concerned. It can then use the catenary to travel out to a suburban area where once again, the vehicle can disconnect and revert to diesel/battery propulsion as desired. The vehicles in service in 2021-2022 can achieve all of the benefits of that basic overhead power facility, and the travelling public and local residents experience immediate discernible benefits.

Initiatives are working with the hypothesis that the hydrogen-fuelled engine power source is to be the locomotive force in the future for public service vehicles. It is perhaps pertinent to speculate where that transition will occur, the timescale involved, and the capital cost.

London claims 50% of all bus mile journeys in Great Britain (7) and boasts 4% of its fleet as electric and 40% diesel hybrid. That compares with the rest of Great Britain and its 94% to 100% (Wales) diesel fleet, of which 3% (outside London) is hybrid - the bulk of which are in England. (9) Aberdeen, it seems, has the starting initiative to bring hydrogen-powered buses with the substantial capital expense that is entailed because 'they' deem to afford it. 'They,' being the people of Aberdeen. It is a small but significant step, and only time will deliver a meaningful cost-benefit analysis. The initiative represents less than 1% of the bus fleet in Great Britain.

What then of Huddersfield, Rotherham, Doncaster, Leeds, Wakefield, Cardiff, Manchester, Liverpool and further north, Newcastle, Dundee and Glasgow? These places have differing circumstances. Their diesel and hybrid vehicles are collectively large in number (in excess of 37,000 buses [excluding coaches]). They comprise the older stock fleets whilst producing an aggravated environmental detriment in close urban communities due to their service vehicle technology age. Of those vehicles, just 31% are Euro VI standard diesel engine vehicles. (8)(9)

These ‘northern’ vehicles do, however, have a huge capital advantage; they cover less mileage and persistent ‘stop-go’ journeys than their south of Watford counterparts. That wear and tear advantage provides for the improved longevity of the chassis, sprung suspension and thereby the life of the vehicle overall. Retrofitting would extend that fleet life further by, say, ten or more years. That factor will, in turn, benefit the operator by extending the original capital cost amortisation period to offset the retrofitting expenditure further. The purchase of new replacement vehicles would be deferred where appropriate. Jobs will be created in the retrofitting companies whilst the bus manufacturers concentrate on their new bus production.

With the statistical analysis we have available regarding the service life of vehicles in those northern cities and similar centres of population around the country north of the South Midlands, we can surely appreciate that the retrofitting of those vehicles, as opposed to those in the south of England might present better value for money through capital distribution. That certainly appears to be so and would fit the stated government objectives regarding regional equality.

The vehicles subject to the program may well come from other locations where newer vehicles displace the hybrid and Euro IV vehicles for retrofitting. Lincoln, for instance, is wholly reliant on non-hybrid buses (DfT definition double-decker vehicles) and is therefore ideally placed to receive 24 retrofit hybrid vehicles to operate four through city centre routes with say 8-9 miles of catenary providing inner-city 'in motion' electrical power for diesel-free transit connecting major destinations (station, hospital, secondary schools, housing and work centres) thus improving the environment and providing evidence of the value for capital outlay (sub £20 million infrastructure/£1.1 million bus retrofit -refurbishment).(Table 2 and 3)

Cost is not the only consideration. The historic landscape of those north of Watford cities, which in the 1800s were the centre of the development of public transport in terms of producing the revenue that allowed the Victorian era to be so productive in advancement terms, remains in quantum in those cities today. Those historic buildings deserve better preservation from harmful chemicals in the atmosphere and their citizens cleaner air to breathe.

Most of those cities had a well-developed trolleybus infrastructure in place from the 1930s throughout the succeeding 30 plus years. This was a progressive advance, in the same way that the early 20th century advanced from the horse-drawn vehicle to the hybrid and steam-driven vehicles, to the tram and then to the trolleybus. The rest was a natural, technological progression then, as it is now. That natural progression in economic, technical and, most of all, environmental terms is required today.

The environment within those cities that would adopt the retrofitting of the current fleet of vehicles would benefit immensely. Historic buildings will be saved from decomposer and blackening caused by the exhaust fumes of the current fleet of vehicles. Streets would be preserved rather than demolished and perhaps 'civilisation,' by way of genuine community, could return to live in and revitalise the city centres. During the past century, dispersal to the outer fringes has left the city centre to be a busy hub for commerce during the working day but devoid of domestic life as a base otherwise. Community health and wellbeing will improve dramatically.

One has only to look at the Americas to witness the decimation of the central urban core of domestic living and its replacement by wide, uncrossable streets, lack of paved footpaths and obliteration of public space. The commercial replacement of dwellings by offices and large retail outlets and the demolition of historic facades replaced by plastic and glass images is all too obvious. That, if reversed, would bring urban centres back to life. The urban 'village' could become the 'des res' of the future if the means of transportation was harmonious with environmentally clean air. Walking on the footpaths of 'urbania' would once more become pleasurable.

The flexibility of the hybrid diesel Euro VI bus and the selective routes of the catenary that provides the overhead power source for the UK bus routes, accessed through the retrofit power poles as roof mountings on vehicles, assists in creating a timely, comparatively exhaust-free environment that would be so welcome in those 'north of Watford' cities and towns as well as London. The need to move out into 'the fresh air' of the suburbs would be reduced considerably, if not extinguished altogether. Surely that is worthy of investigation and implementation if found to be so.

Because of the chassis miles available from these established hybrid vehicles, the retrofitting would extend the lives of the PSV fleet by a considerable margin. It would be at a very modest cost compared to the untimely disposal of comparatively environmentally unfriendly vehicles with a view to full replacement by newer 'cleaner' models purely to achieve the same result that retrofitting envisages, yet at a vastly increased cost. That cost would fall back on to the local population in either higher taxes or higher fares, or both, which is counterproductive in the current economic climate. The production lines for Euro VI hybrid buses with in motion electrical charging would also provide a 25-30 year replacement fleet for the vast number of Euro IV and Euro V buses (19,100 vehicles) (8) that are still in service. They surely are the priority vehicles to replace and perhaps upgrade or modify?

"The big problem with the diesel hybrid vehicles operating in urban conurbations is simply that the battery power does not last long enough to complete a full route journey without recourse to the onboard fossil fuel engine cutting in to provide a recharging electrical source. That was the situation over a century ago."

Roadside power energy stands provide 'supercharge' battery charging that delay vehicle use and thereby earning power, coupled with the unacceptable disadvantage to passenger transportation. The system is also expensive to install and maintain. It has an unacceptable footprint and reduces the already expensive battery life of the vehicles with the relatively frequent attendant replacement battery costs every few years. New Zealand is a prime example of battery power inadequacy in PSVs that fail to negotiate inclines efficiently.(3)

By retrofitting the hybrid vehicle fleet and installing a relatively low-cost catenary infrastructure on selected parts of the route coupled with incorporating a self-coupling and detaching power connection mechanism, in association with that used by traditional trolleybuses, the environmental detriment can be eliminated.

The hybrid PSV fleet will retain its flexibility. Journeys can extend well beyond the urban environment where the constant stop-go, 4 to 7 miles an hour, progression of the vehicle can be elevated to 40 or 60 miles an hour and beyond in suburban and rural areas. This maintains the convenience of longer bus routes encompassing villages and enticing people to use the public transport system rather than the car. Travel will once again become experientially desirable.

A secondary benefit arising from the installation of a catenary system is that the kerbside support posts can be dual purpose providing low-cost modern street lighting, which will further benefit the local economy in energy-saving whilst improving street safety at night on the same structure.

Chapter 10.

A Comparison with Rail

It is at this point probably appropriate to return to the advance of the railways concentrating in particular on 'post Beeching' and a nationalised British Rail system. The current administration incorporates the original two national, if not 'nationalised,' entities: Network Rail and Railtrack. One was created to look after timetabling for train services and associated administration and the other the infrastructure that the trains travel on and are powered by, albeit in a currently revised format. Both were pseudo-privatised government owned entities.

The goods and passenger carrier network comprised a series of independent companies and franchised operations. (23) They operate within a structure dictated by the national entities. (Post-COVID, this arrangement is being revamped as Great British Railways; however, it does not affect the point being made.) By and large, the system has witnessed a huge amount of private capital injection spent on the UK rail network since it was 'denationalised.' That has certainly witnessed an improvement in rolling stock quality in a lot of locations. The question may well be asked; is that advance in rolling stock quality and technology universal throughout the UK?

The answer is certainly not straightforward. However, it is generally clear from the evidence that the 'north of Watford' rolling stock is in excess of 30 years old. The locomotive power is predominantly diesel orientated compared to 'south of Watford,' which is virtually all-electric, apart perhaps, from the southwest rail links to Cornwall and Devon that have yet to be upgraded.

The priority for expenditure on electrification is possibly determined by passenger numbers carried, or the number of trains operated per hour or similar operational density statistics. If that is so, then those 'sound fiscal criteria' apply very much to the detriment of the atmosphere created by the bus services operating in those northern areas mentioned.

Despite the fact that there are high-density populated areas, notably where the trains and buses terminate, those urban areas have largely been placed at the rear of modernisation plans. The result is the attendant concentration of exhaust gases from the diesel engine modes of transport occurring primarily in city centres. That factor is exacerbated by the fact that few east-west routes on the railway mainline 'intercity' rail links have been electrified, let alone expanded to meet modern needs.

The emphasis has been, and remains, on the expansion of the motorway and major road network. Indeed, it was government policy for the railways in particular not to be electrified and that the diesel network be expanded or at least maintained, until relatively recently. Maybe it was fuel prices that determined the bias, which means perhaps that the current collapse of oil prices internationally could do the same again. It will be a short-sighted excuse if it does hamper electrification and the speed of that transition.

Much of the rail network passenger-carrying inventory is leased, which means a lack of ownership on behalf of the rail operators. Born of that is a lack of ability to negotiate from a position of strength. Leasing means that the operators' inventory is 'current account' governed rather than capital value-based, making it wholly dependent upon customer revenues to survive. That is a questionable situation regarding the best option for continuity of service standards in the long run.

Rented (leased) property does give some flexibility where manoeuvrability (change of franchise owner/operator) is required in the medium term. However, is it appropriate to be applied to rail rolling stock or to Public Service Vehicle (PSV) bus operation where the underlying expectation is for the vehicles to be in operation for 25 plus years and the capital expenditure is directed to the 'right to operate,' rather than 'the means of operation' and financial security? Is there a model that will be more appropriate to PSV operations?

With reference to the 'northern' rail system, the leasing of rolling stock is well reflected nationally in the United Kingdom by the impoverished state of Northern Rail and Aviva Trains operating across the middle of the country.

The West Midlands Central and East Midlands rail operators display a well-documented disparity of rolling stock age and quality which deteriorates the further north one analyses and the more rural the service becomes. The time taken to implement electrification, quite apart from the cost, is an inherent factor that has to be acknowledged in environmental terms. Perhaps the subject can now be dealt with in a more pragmatic manner.

Lincoln Chamber of Commerce conducted a valuable analysis of the transportation system some twenty years ago when first considering the cost-effective aspects of 'park and ride' for the city. Attention was focused on the rail network operating into the City of Lincoln. Particular attention was paid to rail usage of the three trains serving Lincoln from the north and west of the city each morning and leaving each evening, carrying just a handful (12 to 16) of passengers and focusing on the small town of Saxilby (population circa 4,000) just a few of miles from the city centre.

The trains operating on that system were old then (and still operate on that route in 2021) and, whilst clean, were, in the winter, cold and draughty commensurate with the stations that they ran from, Retford, Gainsborough, Saxilby.

Albeit that these were well-timed, punctual and fairly speedy diesel train services, they were not, perhaps, attractive to potential passengers. Those that lived in Saxilby and the other similar rail connected towns, who worked in Lincoln, preferred to use their cars. The bus services were also poorly used in comparison with conurbations south of Lincoln. The major exception was North Hykeham, a town in its own right, which has only recently witnessed train usage increase with the advent of a large car park adjacent to the local station to facilitate commuter travel from the new estates into Lincoln City.

Commuters simply failed to be enticed away from modest traffic jams, which also had their detrimental effect on the bus services, on to the speedy train service which would have avoided that aggravation.

As chairman of the transportation committee of the Chamber of Commerce at that time, I often reflected on the ease of provision of cycle racks at places of work and in senior schools in the 1950s. The contemporary attempts of society to replicate that numerical facility in respect of car parking spaces in the 1970s and '80s as private vehicle use became more prevalent, failed miserably.

Modern transport hubs with ample, well-priced car parking spaces are certainly part of the answer for the railway networks. Should we not consider 60-80 people travelling in 40-60 cars who could be transferred to an 80 passenger modest footprint electrically powered trolleybus style PSV? That vehicle would carry them the 4 to 8-mile commute into a major city on the existing road network from a 'Park and Ride' facility in the shape of a full community development, such as a residential estate.

How much more space would be released, and how profoundly would the environment be improved with the application of that facility? Bus stops are certainly cheaper than the installation of railway stations. The same cost principle applies to tram stations and the major infrastructure impediment that they create quite apart from the capital cost of their implementation. (11) Cost-benefit analysis reports from Manchester(12) and Sheffield(13) prove interesting examples when subject to comparative evaluation of the tram system in operation and the infrastructure amortised cost relative to a break-even point. They are financial drains on scarce resources but arguably provide an essential service once established, at a cost.

It is perhaps time to allow for a vested interest to bid for the funding of electrification of rail links for longer journeys. Consideration is also needed regarding road links and public service vehicles on intermediate journeys for those travelling into work and service centres, as well as sightseers and shoppers.

The problem always seems to be a disproportionate allocation of resources to those with the loudest voice or the most informed barrister. It is perhaps time for the expertise of engineers to be taken into account and projects costed out, rather than has been in recent times; a budget prepared by a master contractor and a free-for-all to bid for the work, ignoring the engineering arguments that actually make it happen.

The tendering system has fallen into disrepute as many participants use the rules rather than the facts to gain access to resources - financial and otherwise. Large organisations bid and subcontract aspects of their fixed-price contracts, often browbeating smaller subcontractors into a nightmare operation of delayed payments and overbearing authoritarian administration. The primary contract for road and rail construction is a prime example of chaos arising from that example of 'major contractor' default concerning the Criterion Group.

History can provide a guide. The indications are that over the next 20 to 25 years, as the development of hydrogen power, or perhaps another yet to be discovered entity, takes place, the efficient use of electricity should be the emphasis for a 15-to-20-year strategy that has an immediate benefit to society as a whole. It should be centrally generated rather than inefficient hybrid generated or based on unreliable battery power.

In the meantime, surely, we must use what we have available to the best advantage. It is not possible to change the world overnight, or even, with current production methods, in ten years. However, by adapting and retrofitting, modernising and upgrading existing vehicles and their power sources, time can be used efficiently. The environmental standards achieved in a shorter time period will benefit all. The workforce and facilities are available.

In 1946, the London Passenger Transport Board decided to remove all remaining trams in London and replace them with diesel buses. This change was really the start of the transfer from electricity-powered vehicles to diesel-powered vehicles, which later encroached on the trolleybus system. Where London led, others followed. Perhaps it is time for the North to be in the vanguard once again, rather than the guard's van. Public transport started north of Watford.

Another consideration regarding electrically powered vehicles was the nationalisation in 1948 of electricity suppliers. It meant that those who provided cheap electricity could no longer do so, even where they owned their own local power company. That made running electric vehicles more expensive, in favour of diesel, which at that stage accounted for 8% of the fleet operating costs. Why would it be so difficult to allow the PSV operator to provide the power source required?

Competition creates advancement. Need is said to be the mother of invention, and we now have a most pressing need for that inventive power and a listening ear to carry it forward.

Ironically, the cost of diesel fuel rose in line with oil prices and soon became a massive 24% of running costs for the diesel bus fleets of the UK in the 1970s, increasing the cost of service provision.

Chapter 11.

A Modern Analysis of an Age-Old Problem

The 2013 INTERREG Central Europe Project TROLLEY (1) is an extensive paper full of research covering seven European cities; Salzburg in Austria, Gdynia in Poland, Leipzig and Eberswald in Germany, Brno in the Czech Republic, Szeged in Hungary and Parma in Italy. The aim of the paper is to promote bus systems with zero emissions through the trolleybus principle. Needless to state, Europe has a substantial trolleybus network from which to draw information and experience. With that in mind, it is a paper to take note of.

The full title of the paper is: *Take-up guide for the Replacement of urban Diesel buses by Trolleybuses.*(1) Comprising 304 pages, this document can be a daunting but worthwhile read. Please note the emphasis on ….'urban diesel buses' in the paper's title.

The paper admits that in many areas, and of course the UK is a prime example, the trolley bus became less than fashionable due to the vehicles' perceived inflexibility, the overhead wires, and perhaps because they had been around for such a long time, outdated technology.

Therefore, diesel-powered buses took on the mantle of a flexible and comparatively economical vehicle to run.

Much of the decision-making process in the United Kingdom was arguably politically orientated in support of the nationalised bus industry. It ignored and excluded the bodybuilding firms involved with a trolley bus fleet, coupled with the regulation of much of the bus network, further stultifying commercial competition in a very busy marketplace. There was a perceived oversupply of producers/manufacturers to the bus industry at that time.

The dynamics are very different on the continent. The 2013 TROLLEY paper (1) exemplifies the growth of the existing networks as they continue to play an integral role in keeping the air clean through green transport systems employing the trolleybus and the battery electric bus principles. The In Motion Charging project in the UK is able to capitalise on the extensive knowledge that prevails in Europe regarding the operation of electric vehicles and particularly the trolleybus and its development.

The biggest challenge to the operation of trolleybuses or any public service vehicle system is the growth of personal ownership of the motorcar, and here, the United Kingdom excels. The DfT analysis tables clearly show continuing growth in motor-vehicle ownership of almost 25% over the last ten years. (20) With the advent of electric vehicles and the environmental pressure to go 'green,' coupled with the substantial increase in the purchase price of electric vehicles over a conventional fossil fuel competitor, one has to ask whether that growth in personal car ownership will continue. Conversely, will the introduction or expansion of an environmentally friendly and efficient public service transport system reverse the personal car ownership trend?

The European paper(1) makes the same observation. Many European countries, especially those that were under communist Russian influence for so long, are becoming more affluent and engaging with the private motorcar. This is creating congestion and the environmental hazards associated with it in urban areas. Asia has exactly the same problem. Public transport usage is in decline.

The trolleybus, the paper states, is not the answer to every problem. Nevertheless, it can be the essential core of a large conurbation transport system if properly placed to provide service to the areas of the large residential population who require to be linked to their respective service and manufacturing areas. There is, however, no overnight solution.

The report did have an effect in the city of Leeds, resulting in an application for a trolleybus-style installation along the lines of the 2013 report concerning the installation of a fixed-route, catenary-powered system based on single or double articulated vehicles. Other European cities have installed systems involving 24 m long vehicles travelling over purpose-built road networks, akin to a track, but employing the trolleybus principles in the power supply and propulsion elements that the 2013 paper promoted.

The paper also promoted the superbus on the basis of a non-rail-based tram-style system using articulated three carriage vehicles across a city centre. There was a great similarity between the general principle and application proposed by Leeds to the Manchester, Sheffield, and Birmingham 'tram' systems submitted. These are very large UK cities and have similarities with the cities of Europe that this report is based upon. The Leeds project was rejected by the powers that be.

Currently, trolleybus systems are employed in over 310 cities in 56 countries. There are currently some 44,000 trolleybuses in operation around the world. In many locations, it is a very modern means of transporting the public. The TROLLEY research states that 45 new trolleybus-based systems were created in the world in the 12 years prior to the 2013 report; 27 of those were in Europe.(1)

The trolleybus principle has proved itself to be the most reliable electric form of vehicle. It is currently the most common electric on-road urban public transport vehicle worldwide. Therefore, there is every opportunity to take the known technology and experience gathered over the last century of trolleybus operation to achieve the environmental objectives of the European communities transport White Paper: *to reduce Public Service Vehicle emissions by 60% before 2050.*

Given our current capacity to manufacture and replace existing vehicles, that timescale is realistic if it is acknowledged that it is progressively achievable. The emphasis must be on progressively. There is no overnight solution. As explained elsewhere in this book, there are 15,000 sub-Euro VI diesel engine vehicles operating in the UK alone. Manufacturing 15,000 new buses to replace that huge portion of the total circa 38,000 bus fleet cannot be achieved overnight. The other 23,000 vehicles will need upgrading at some stage and possibly concurrent with their counterparts.

As the EU 2013 paper(1) observes, the trolleybus is the only heavy-duty public transport vehicle available in versions of 12 m, 18 m, and 24 m which can not only equal the operational performance of diesel fossil fuel buses but also exceed them. A diesel bus can remain in service until it has to be refuelled, whereas a trolleybus can remain in service indefinitely attached to its catenary as a permanent power source. The trolleybus vehicle provides the optimum fuel and time-in-service efficient mode of transport, bar none. Battery Electric Buses (BEBs) do not provide the level of operational efficiency per vehicle that either diesel or trolleybus operation provides on major bus routes with heavy ridership in either efficiency or economic performance.(10) That factor is repeated elsewhere in other comparative research papers.(1)(2)(22)

Given the modern, updated version of the trolleybus as an In Motion Charging facility and the incorporation of the onboard battery element making the vehicle capable of several miles of unsupported electric charge away from the catenary, the trolleybus has morphed into a more flexible means of transport. It is an All Service Vehicle (ASV). (5) Now that the flexibility of operation is achieved, moving away from the perceived inflexibility of the overhead cables to a more up-to-date recognition that the catenary provides In Motion Charging for the onboard battery facility, the concept takes a more acceptable role. The nature of the vehicle has fundamentally changed.

The analysis paper's conclusion is that society should acknowledge and take advantage of that change. Bus designers have to determine the efficiency of bus autonomy, connected or disconnected from direct power, and the battery pack weight incumbent in the route requirement for energy storage. Similarly, the facility within the vehicle to store power and determine the range of travel also exponentially detracts from its passenger-carrying capability and journey duration. There is a maximum vehicle load that the chassis will endure and the time over which that endurance will run out (wear and tear - vehicle life). Here one draws a comparator with the American experience and the research paper on automated vehicles using a 12-year life duration of each vehicle against the author's analysis and the UK expected life duration of double that period or more. (2)

A meaningful time-cost-benefit ratio must surely be taken into account. Too much battery weight is a vital component of that ratio. Several commentators refer to the pure battery bus being tailor-made to service certain route requirements. This must reflect in the flexibility of fleet operations and, thereby, the number of vehicles in the fleet to allow for route management over time. (10)

The 2013 paper(1) does consider the hybrid electric bus. It precludes any fossil fuel involvement but replaces the diesel engine, biofuel, hydrogen element needed to generate electricity to the provision of that electricity for the onboard batteries from a catenary system operating as a trolley bus. It then disengages to continue the route on battery power until it re-engages with the catenary power to continue. In that way, the report points out, one eliminates the disadvantages of the overhead connecting system and its inflexibility whilst also overcoming the short-range duration of onboard batteries as a means of propulsion. The overhead catenary can be reduced in many instances to 50% of the route requirement.

The electric motor is in itself an optimal mechanism for transferring power by delivering an energy efficiency of some 40-45% against any mechanical transmission, which achieves an estimated 25% by comparison. Energy efficiency is therefore achieved if placed in the hands of an electric vehicle compared to its diesel compatriot. (1 page 11)

Since 2005, many extended articulated trolleybus vehicles have been developed and used within Europe. This provided very valuable in-service information that was used by the 2013 EU report (1) to compare efficiency and service, coupled with a level of operational acceptability to the general public with overwhelming acclaim.

The mode of transport is very specific and, because it is the driving topic within this report, is treated appropriately when considering the general use of battery electric vehicles and the trolleybus in motion charging comparatives. In many instances, they are just not comparable. The supertram concept trolleybus vehicle has a very well defined and specific location for its potential operation. The Cambridge Park and Ride facility comes to mind as a potentially suitable system for consideration. However, the general principles and the comparisons drawn within this report give sucker to the general discussion regarding the replacement of diesel vehicles with electric vehicles at one time or another.

From the data produced and bearing in mind the experiential evidence rather than academic assumptions used in the report, the most efficient means of driving an electric vehicle is from an overhead catenary. Any desired flexibility for the vehicle is achieved by onboard battery stowage back up to provide for journeys away from the overhead power supply. The author's observation is that a hybrid Euro VI diesel range onboard facility should not be ruled out where long and undulation topography are involved.

The main emphasis of this report (1) is focused on the powertrain principle. However, it does give time to explain that the number of trolleybuses required during peak periods and the timeline and routing of those lines, plus the environmental conflict with other road-using vehicles sharing the same route, must be analysed at the planning stage. Similarly, the type of vehicle and its performance characteristics with reference to the form of electrical energy charging and onboard storage management systems need to be considered, and the cost of placement of that infrastructure, its expected life cycle and anticipated efficiency timeline.

The power supply network is an essential component of a complex time-cost-benefit analysis required to identify the routes that need to be interconnected, coupled with where and how those interconnections should be provided for by way of on route battery charging stations if appropriate. The most economical installation has to exercise consideration for the operational, technical, and urban factors that will influence potential journey switches and for any catenary interlinking. Any preliminary assessment of the required energy supply system needs to blend in naturally with the urban landscape and not interfere with historic buildings, street signage or potential festival street decorations.

The positive element inherent in the introduction of a modern trolleybus system incorporating the freedom to leave the catenary and use onboard power for part of its journey is that local communities can thrive, and new communities develop in the knowledge that the air around them is clean and healthy. The 2013 EU report explains in some depth the variation and dual purpose that can be achieved from the catenary infrastructure arrangements.

A catenary can facilitate street lighting, public address systems and similar supplementary, socially acceptable improvements to the local environment. The cost of infrastructure can vary between €1 million and €20 million per kilometre for implementation. Much more realistic and modest by comparison figures are quoted elsewhere in regard to the analysis of potential costs. In the 2013 report, (1) the higher figure of that assessment relates to the super train and the demolition of existing structures in order to place an up-to-date and designated route for these 80 feet (24 metre) long articulated vehicles.

Part of the problem in the United Kingdom, as Leeds Transport Authority found to its cost, is the Transport Act 1985. In this modern day, with modern technology, it is certainly outmoded in its requirements and time-consuming protocols regarding planning permissions and bureaucratic requirements. They would perhaps be better placed at a local level within the applying local authority than with central government and its limited knowledge of local affairs.

A more accommodating compromise arrangement between local and central government could be incorporated in a revision to that original act or its replacement altogether, particularly with reference to the *Bus Back Better* UK government initiative. (5) Planning consent is a time-consuming element, and time is against the environmental and *Bus Back Better* defined clock.

Qualitative comparisons of various bus systems (EU definitions)[single deck units] Tables 5.6 - 5.7 and 5.8 EU 2013

advantages of electric operation

	table 1 of 6	**diesel / diesel hybri**	**battery bus no route charging**	**battery bus route charging**	**batterybus charging at random stops**	**trolleybus full catenary**	**electro - hybrid partial catenary**	**fuel cell hybrid**	**induction power supply in motion**
2	low noise		yes	yes	yes	yes	yes	compressors	yes
3	no vibration		yes	yes	yes	yes	yes	yes	yes
4	lpowerful accelaration		yes	yes	yes	yes	yes	yes	yes
5	Low-maintenance, long-life electrical components		yes	yes	yes	yes	yes	yes	yes
6	Low operating costs (in terms of energy consumption) and a high level of independence from energy suppliers		yes	yes	yes	yes	yes	yes	yes
7	A clearly visible, high-quality service					yes	sectional		yes

comparison of disadvantages of electric operation

	table 2 of 6	**diesel / diesel hybri**	**battery bus no route charging**	**battery bus route charging**	**batterybus charging at random stops**	**trolleybus full catenary**	**electro - hybrid partial catenary**	**fuel cell hybrid**	**induction power supply in motion**
9	Sufficient energy reserves in the event of a fault in the power supply					It is assumed that there is an independent from the basic technology -an on board energy storage unit			
10	Low impact on the cityscape thanks to use of customised materials and specific designs for overhead lines, masts, charging stations, etc.		n/a	charging stations	charging stations	a full catenary	partial route catenary	n/a	enbeded in the road structure

operation similarity to diesel buses

	table 3 of 6	**diesel / diesel hybri**	**battery bus no route charging**	**battery bus route charging**	**batterybus charging at random stops**	**trolleybus full catenary**	**electro - hybrid partial catenary**	**fuel cell hybrid**	**induction power supply in motion**
11	Availability of different vehicle sizes depending on purpose or demand		currently only 12metre	currently only minibus	currently only 12metre	yes	yes	yes	currently only trial vehicle
12	performance up to 20hrs per day early am til midnight		10 hr	At Turin 12-13 hour	target	yes	yes	12hr	target
13	range over 400km/day with no complete charge				target	yes	yes	250 km	target
14	simple and fast refueling or charging		central overnight charging	central overnight charging	target	yes	yes	special infrastructure	target
15	easy operation unimpaired by other traffic		yes	yes	yes	yes	yes	yes	separate right of way
16	Dwell times by passenger demand not extended by technical needs		yes	minimal break time	charging 30sec. Dwell time 20 sec	yes	yes	yes	yes
17	sufficient power supply to auxilliary units. Air con, lights, doors etc		target	target	target	yes	yes	yes	target
18	simple handling by driving staff. No distractions or special knowledge required		yes	yes	continuous suprvision status of charge	yes	yes	yes	target

investment costs similar to today's systems									
	table 4 of 6	diesel / diesel hybri	battery bus no route charging	battery bus route charging	batterybus charging at random stops	trolleybus full catenary	electro - hybrid partial catenary	fuel cell hybrid	induction power supply in motion
19	vehicle and compoent costs comparison to straight diesel euru V					coct 1.5 to 2 times more than Euro V diesel		cannot be assumed	cannot be assumed
20	competition for purchase		assumed	assumed	one supplier 2013	yes	yes	two suppliers 2013	monopoly
21	infrastructure en route /at stops/ along the route		no	charging station	charging station	catanary	catanary - sectonal	no	pow er supply in lane suface
22	infrastructure in depot / workshop		electric components	electric components	electric components	electric components + catenary equip	electric components + catenary equip	electric components and hydrogen	electric components
23	future development and availability of spares		assumed	assumed	cannot be assumed	yes	yes	cannot be assumed	cannot be assumed

Cost of operation and mainainance									
	table 5 of 6	diesel / diesel hybri	battery bus no route charging	battery bus route charging	batterybus charging at random stops	trolleybus full catenary	electro - hybrid partial catenary	fuel cell hybrid	induction power supply in motion
24	lower cost of energy consumption		expected	expected	expected	expected	expected	expected	expected
25	independancy from price flucuations on worldwide market		expected	expected	expected	expected	expected	expected	expected
26	less wear and tear by electrical components		expected	expected	expected	expected	expected	expected	expected
27	longer life cycle of vehicle and components		expected	expected	expected	yes	expected	expected	expected
28	operating speed up to 60 miles per hour (95.7 km/hr)		50 mph - (80 kph)	50 mph - (80 kph)	50 mph - (80 kph)	yes	yes	expected	expected

conclusion of evaluation by comparison with diesel bus operation									
	table 6 of 6	diesel / diesel hybrid	battery bus no route charging	battery bus route charging	batterybus charging at random stops	trolleybus full catenary	electro - hybrid partial catenary	fuel cell hybrid	induction power supply in motion
(number of criteria per evaluation level)									
	improvement by comparison	0	10	10	10	12	12	10	12
	equal by comparison	28	13	10	9	12	12	11	8
	additional effort /cost by comparison	0	2	5	6	4	4	2	4
	Significant degradation/no clear data	0	3	3	3	0	0	5	4
	percentage comparison + / -								
plus	improvement by comparison	0	35.71%	35.71%	35.71%	42.86%	42.86%	35.71%	42.86%
equal	equal by comparison	100.00%	46.43%	35.71%	32.14%	42.86%	42.86%	39.29%	28.57%
minus	additional effort /cost by comparison	0	7.14%	17.86%	21.43%	14.29%	14.29%	7.14%	14.29%
minus	Significant degradation/no clear data	0	10.71%	10.71%	10.71%	0.00%	0.00%	17.86%	14.29%

The above report from pages 79 and 80 of the EU document(1) is helpful in that it also has a cost calculation model on pages 69 of that extensive 304-page report.(Tables 5.6, 5.7 and 5.8) What must be noted, however, is that the diesel comparisons are all based upon Euro V engine powered vehicles as against the significantly improved Euro VI engined successor vehicles.

The calculation model was apparently used with some success by the city of Esslingen am Neckar in Germany which, with a 90,000 population and a population density of 2,000 inhabitants per square kilometre (5,200 per square mile), discussing the purchasing power per resident coupled with an unemployment rate, which provides the basis for its assessment. The ideal trolleybus network size for the city was estimated to be 20.5 km (12.738 miles). The result of using the model indicated that the cost to build the trolleybuses system would be €23.9 million. €16 million of that would be the network cost which led to an anticipated €800,000 per kilometre of system implementation or €1,287,200 per mile. (£1,100,000 per mile). This was a far more comfortable figure to work with than the parameter costings stated in the earlier indicative cost of the tram-style PowerTrain project. This indicates the significance of simplicity in infrastructure terms.

The analysis is said to include an assessment of the capital investment cost to design, build and commission the electrical power network, the electric bus depot and the electric bus control centre. There is also an assessment of the capital investment costs for the acquisition of the trolleybuses (standard size 33 foot [10 metre] vehicles) and the required maintenance system tools, equipment, and vehicles, coupled with the assessment of the potential saving expenses on a recurring operational budget. NB The vehicles compared are invariably single-decker ('coach' by DfT definition).

This element of the 2013 EU trolley bus report is an extremely useful and helpful indicator for local authority transport departments when assessing the budgetary requirements for any installation designated to replace a fossil fuel (diesel) fleet of Public Service Vehicles (PSVs). It indicates the control needed to replace diesel engine vehicles with electrically powered vehicles and the procurement potential for their own local environment. The need for a local zealot appears to be a primary requirement for a successful outcome. Therefore, it is to be hoped that this publication helps in the process of creating zealots galore.

Where there is a wish to act speedily and reduce the ecological and environmental footprint of a predominantly diesel bus service, caution has to be exhibited. There is an overriding need to recognise the advantage of a structured, forward-looking though temporary, perhaps ten years, transition period required to prepare for future technologically advanced solutions to the public service vehicle environment. The hybrid diesel vehicle, even if manufactured today with the Euro VI diesel engine, has to be considered within this, warts and all. The incorporation of a supplementary overhead power system provided by a catenary has to be a contender for the prompt advancement of the move to more environmentally friendly urban operation using the trolley bus system to mitigate the harmful effects of diesel propulsion in inner-city and urban environments.

Some 310 cities worldwide, 150 of which are stated to be in Europe, currently operate trolleybus systems. Some of those, such as Vancouver in Canada, have been operating the system continuously for over 70 years. Many European trolleybus systems far exceed 30 years of history of operation. The fundamental change in the nature of the mode of recent trolleybus operations gives it a dimension that history could fail to foresee in its flexibility of operation with onboard battery incorporation.

The report goes on to discuss the sources of electricity and how the cost of production of electricity impacts on the diesel/electric vehicle debate. Suffice it to say that the report is broad-based and sourced from factual rather than ephemeral information, which leads it to have authority in the assessment process. That will assist those who decide the nature of public service vehicle provision in towns and cities for the next 20 to 30 years. That is certainly the timescale, the minimum timescale, that needs to be used in any current assessment under the UK Department for Transport's *Bus Back Better* initiative in 2021 and beyond. (5)

The following is an extract from an 'informed paper' (14) on transportation.

There are apparently "six ways to **improve public transport** which are simple and inexpensive" and recommended for making public transport more attractive to customers.

1. The first is to speed up vehicle boarding. How do you do that? The only way, the report suggests, is to **use pre-boarding ticket sales**. Not a bad idea. It is employed in many areas of the world.

2. Use **low-level boarding** or low floor vehicles. This, of course, incorporates improving vehicle design "to pack people in" whilst simultaneously (and contentiously) accommodating those with disabilities, pushchairs and young children.

3. A further recommendation is to **optimise 'bus stop' design and location**, in other words, where the passengers gather to embark (or disembark). The report suggests reducing the number of 'stops' or *carefully locating the stops* on the street where people will want to use them. We have witnessed the infrastructure change to accommodate raised pavements at bus stops and kneeling buses. These ideas were added to the infrastructure and vehicle costs and are now widely in place around the UK.

4. **Reduce traffic congestion**. How is that achieved? By coercion which creates resentment? The report recommends providing *dedicated lanes for public transport* (which, of course, means making the road smaller for other modes of transport to use by creating bus lanes or permanent PSV tracks or highways). [Author comment - The impact of the Powertrain project can and often does create the very traffic chaos that the principal is designed to improve.] Another suggestion is to *use regulation and traffic engineers to control traffic*. [That has been a demonstrable failure across the country over the last few years in most of the cities visited. Evidence of this has been the closure

of so-called 'rat runs,' known to locals and emergency service operators, and the recent cycle path, or should that be psychopath, which were solutions to back street traffic flows.]

5. The fifth idea is to give public transport **priority at traffic signals** so that they reduce the time spent waiting. This facility apparently makes customers /passengers happy and reduces costs, the report averred. [More than a few observers would beg to differ. Some interpretations make other road users angry and frustrated when the project is imposed over a minimalist length of highway. There is little or no practical outcome for the PSV or other road users, yet a significant cost to the local authority that cannot be justified, other than by 'tokenism'.]

6. The final report suggestion is to use real-time data to control public transport operations, improve public transport planning, and of course, provide **accurate information to customers**. Display screens in bus shelters – you don't mind waiting if you know the bus is on its way. Yes! Do away with printed timetables? No!

There are a number of papers with variations on this theme.

[The author's submission is number 7, to add to the foregoing. **Keep the footprint of the public service vehicle small** rather than multi-length 'bendy bus' or large platform/chassis single-decker units. Double-decker buses load and unload pretty much as quickly as a single-deck bus because of the time alighting passengers spend forcing their way through the standing crowds on the modern 'standing room only' bendy bus facilities. [The "stand all the way to your destination" vehicles do not appear to be popular with the travelling public.]

Support complementary measures and, of course, complementary transport. Planning for transport in urban areas might include traffic calming. 'Sleeping policeman,' bumps in the road and chicanes impede comfortable public transport vehicle progress and achieve little if anything other than increasing exhaust detriment from fossil fuel vehicles, further polluting the local atmosphere as vehicles accelerate away from the enforced impediment.

Cycle compounds together with **public bicycle systems coupled with parking management** and realistic pricing strategies are also considered within the report.

It is fairly obvious that few if any of the people involved in this planning exercise remember the hordes of cyclists that habitually left places of work (primarily factory units) at much the same time, morning, noon and night. They would 'fight' their way home on broad streets with virtually no vehicles other than bikes on the road to impede their progress except for public transport vehicles whose progress was as impeded by the 'bikes' as they are by the modern car.

That's progress.

The bicycle is promoted as such a spirited technological and personally superior mode of transport to return to. However, we should not ignore the accident statistics that went with that 1950s/60s phenomenon. History teaches us so much - if only the planners would refer to the evidence that it conveys.

The main bus fleet operators include:

Abellio, Arriva, First Group, Go-Ahead Group, National Express, RATP Group, Stagecoach Group, and Transdev. These are the dominant transport groups operating bus fleets in the UK. Have they, and more importantly, their drivers, been consulted or involved in the various suggested principles advised above? Perhaps they will concur. They may provide a broader insight and additional ideas. It is worth asking the drivers and other passenger-facing staff for their observations and opinions.

Chapter 12.

Electrifying and Automating Bus Transit Fleets

This is another interesting technical paper that immediately invokes further thought of Cambridge and the experience of its Park and Ride transit system. It immediately engaged the thought pattern regarding the possibility of something contentious coming into being in the United Kingdom.

The study by Neil Quarles, Kara Kockelman and Moataz Mohamed is extremely interesting and up to date, having been published on the 13th of May 2020. (2) The criteria used in the report are relevant to a broader-based discussion than simply the use of electric-powered buses. It promotes automation as well: driverless vehicles. That broad remit is what makes the paper such a compelling discourse in respect of the dismissal of the diesel bus, the fossil-fuelled engine public service vehicle per se, and the environmentally more acceptable vehicles in the current debate.

The basis for the report is the Capital Metropolitan Transportation Authority bus fleet in Austin, Texas, USA. The authors use US fuel prices of $2 (£1.41) a US gallon (

0.37P/ltr) for diesel fuel to assert that, taking account of the time/cost nature of their charging infrastructure, battery electric buses will become a life-cycle cost, competitive against diesel fleet vehicles by 2030. The speculative element of that statement is offset by actual cost comparisons relating to in-service vehicles used in the observations.

One of the attractions of the report is that it does not hold an inherent bias against diesel fuel operated vehicles. The study merely views public transport provision in a cost-benefit environment, which includes removing the driver from the bus and automating substantial parts of the public service vehicle provision in large cities such as the subject city: Austin, Texas. It concludes that 83% of US passenger trips are conducted in a private car. Austin does not have a significant rail network to complement its road traffic facilities. Therefore, 93% of Austin's public transport journeys are made by bus - predominantly diesel-powered and single-decker (DfT - coach).

Just as in the 2013 EU report (1) on replacing diesel with electric-powered vehicles, there is an emphasis leaning towards powertrains. In the United Kingdom, this probably translates to the Manchester and Sheffield overhead power supplied tram systems and the recent Birmingham experience of battery-powered trams (no catenary) regarding their ability to satisfy a public need within reasonable capital expenditure and the life expectancy of the system. [Cost-benefit analysis over time.]

A quoted third party observation (Lajunan et al.) concludes that hybrid/battery electric buses are favoured with respect to their life-cycle cost/operation when considered with the environmental measures for transit application. However, even when considering different charging methods, pure battery for the electric technology is still favoured over the hybrid alternative onboard power source for battery recharging. The report goes on to discuss other investigations that came to that same conclusion.

Fully automatic electrically powered public transit systems are not new to the United Kingdom. The Docklands Light Railway (DLR) system operating from Tower Gateway and similar stations to the area around London City Airport and Poplar has been an automated system for over 30 years. It remains very successful, apart, perhaps, from its inability to collect all fares due in the early days, as personally noted by the author on his frequent journeys on those routes. One must only assume that the other route suffered from the same penalty. Hopefully, that shortfall in income has been addressed in the more recent past.

The point being made here is that automated systems have to be successful to the point that the majority of, if not all, passengers pay the requisite tariff to use the system that is in place, other than being honour bound to do so. The honour system can only go so far.

By using the city of Austin and that city's Capital Metro bus system as a case study, the report focuses on the inherent foibles or otherwise of that particular location. Nevertheless, the report is wide-ranging, factual and therefore influential in putting one system of public transportation and its propulsion against another. It asks the question, "What cost-benefit criteria affect the budgetary influences that would direct an investigator/assessor to choose one form of propulsion, say diesel, over an alternative, say, battery electric, or vice versa?"

The environmental argument is certain in its outcomes, and the report proves that conclusively. However, it discusses natural gas, which may be considered, along with biofuels, as a replacement for diesel fuel. The conclusion quoted in the subject report (2) is that since natural gas/biofuels result in toxic emissions in a similar way to diesel, even from the most efficient engines available, they would not meet the environmental criteria required in the long term when compared to electric power from natural resources.

Hydrogen fuel cell electric buses have been subject to several pilot programs in the United States, the report points out. The problem appears to be that the source of hydrogen itself generates emissions because it is derived from methane or other gases, the production of which creates carbon dioxide (a greenhouse gas) as a by-product. The conclusion of the *Tong report* (5) quoted here is conclusive in that hydrogen produced from natural gas has a similar emission problem (albeit localised to the hydrogen production plant) to the modern diesel engine. Therefore, it is less environmentally satisfactory as a primary power source even when linked to an electric power plant within the bus.

The report goes on to examine the source of electricity used to charge battery electric powered buses or power them through a catenary. Once more, there is conjecture over how much is produced by hydroelectric power, the sun, wind and so on, against gas and fossil fuel-derived power stations.

Strangely the report does not mention nuclear energy, which the author, along with Rolls-Royce, and other small nuclear reactor electricity generating plant manufacturers, including some in Canada, might draw attention to. There is an alternative power source per se. Why is it ignored? Electricity generation on a city-by-city scale follows on logically from the nuclear electric powerplants used in nuclear submarines and other naval vessels that have evolved from the late 1950s to eminently safe reactors in the 2020s. Britain could once again lead the world.

This interesting report does discuss hybrid electric vehicles that could be considered, notwithstanding the emissions that will occur away from large conurbations under diesel power. However, it is acknowledged as an acceptable, viable, stop-gap vehicle to the pure fossil fuel bus by supplementing the power charging facilities invariably required by the electric-powered bus. It charges the batteries on the move.

Returning to the Battery Electric Bus (BEB), the report acknowledges that there are two means by which onboard batteries can be charged: overnight or en route. The study suggests that the BEB charging infrastructure differed according to each transit operator's unique characteristics. For example, the route profile, timetables, topography, geography, weather conditions and so on, quite apart from the cost-benefit attributes associated with each BEB technology, all contribute to the ultimate decision-making process.

The American report (2) analyses fuel expenses over a 12-year period of operation of each vehicle, whereas studies in Europe, and the author, tend to be more pragmatic regarding vehicle life. It should extend to 20 to 25 years at least for a UK fleet vehicle to justify the vehicle's capital cost alone. Refurbishment may occur at the 12 to 15-year point to upgrade the vehicle.

The capital cost of each vehicle is assumed in the report to be between $300,000 and $400,000 [£224,719] on the assumption that a government grant might be available to reduce the larger to the smaller amount. The UK analysis ignores subsidies for comparative purposes. As the tables of UK costings clearly show, the capital cost of vehicles differs tremendously (as might quality and life expectancy). However, the assumption of 200 kg (31.5 stone [440 lb] assumption for a battery pack for a battery electric bus is comparable). NOTE - this report is again dealing with single-decker [DfT coach] vehicles. Double-decker buses (DfT bus) carry ballast as a centre of gravity lowering mechanism, which is merely replaced by the weight of the hybrid vehicle's batteries.

Suffice it to say that it is not the purpose of this chapter to go into the USA report in detail. That can be achieved by the reader from the report itself. (2) However, the analysis does include the cost of drivers, their training and employment costs which, similarly, is taken over a 12-year period. The rationale for the report comes out at this stage when it examines the cost of automating the driving of the vehicle and the cost-saving through not employing a human being to fulfil that function. That is the principle of the DLA in London. Some would say that it works to a level that makes it pertinent to this analysis and therefore worthy of inclusion, particularly with the Cambridge Park and Ride in mind.

The *Sustainability 2020*(2) report incorporates a fair number of assumptions, many of which are drawn from large metropolitan transit authorities, all of which are in the USA.

Strikingly, the direct cost analysis concludes that battery electric buses (BEB's) are <u>not yet</u> totally cost comparative with diesel-powered buses. However, fully automated buses (in other words, those without a driver) should become competitive immediately when they are brought into service. The only caveat to that, the report states, is that electric bus prices are falling, largely due to the reduced cost of producing the onboard batteries.

Consequently, there is a suggestion that within the next "couple of years," the original assumption/conclusion of the analysis could be reversed. It must, of course, be remembered that the situation in the USA with regard to fossil fuel availability, underlying costs and perhaps even the production of electricity may vary considerably to the United Kingdom and Europe.

Where perhaps the conclusions of the report do leave the door open is in respect of justifying the intermittent use of hybrid combined 'In Motion Charging' (IMC) battery-operated vehicles drawing power for at least part of the journey from an overhead electrical source (a catenary). The combined benefits of the three elements involved in that curious combination could well bridge the production gap that all the reports acknowledge in their fulsome analyses.

Chapter 13.

A Canadian Perspective

As a visitor who is personally acquainted with the Canadian cities of Vancouver and Toronto public transport systems, albeit that they are 3000 miles apart, one gets a sense of two citizen-orientated public transport service operations in equal measure.

Vancouver has operated a trolley bus public transport system for over 70 years with great success and acknowledges the long-term benefits of the buses that they operate. However, using Department for Transport definitions, we should call the vehicles coaches because they are all single-decker units, and it is believed always have been.

The public transport system in Toronto is diverse in its application. It encompasses various means of transportation by combining the bus transit fleet and the Metro underground railway system. It is commonplace to purchase a ticket on either bus or Metro and use it interchangeably to complete a journey. Toronto's bus system is all single decker (DfT coach). The city does have double-decker trains on the Go-train network. However, that system is an entirely separate entity and does not have connecting ticketing facilities.

Toronto Transit's public service operates underground trains (Metro), and their surface transport, which interconnect by means of a ticketing mechanism. Passengers buy tokens that are exchanged for a ticket, and that ticket can, in turn, be transferred from route service to route service until the passenger reaches their destination. A bus from Chinatown to downtown, transfer to the metro and then at the metro terminus transfer to a bus into North York is achieved on one ticket purchase. Like transport for London, there are travel zones with set tariffs.

The organisational structure of the Toronto public transport system melds very well with the aspirations of the *Bus Back Better* paper recently issued by the UK government's Department for Transport (DfT). (5)

The foregoing leads one to have great confidence in the robustness of a recent paper which compares the service aspirations of the two comparable Canadian cities in terms of the potential acquisition of an electric vehicle fleet for each of those two public transport service operators. The big problem, if indeed, problem is the correct word - dilemma may be more appropriate but still seems a little strange to use - is that the subject matter of electric buses and particularly the cost-benefit thereof is multifaceted. What initially seems to be expensive amortises that expense over time and therefore seemingly provides better value for money in the longer term.

Against that, there are short-term advantages which were perceived with the introduction of the diesel bus in the 1960s, so many years ago. The diesel-engined vehicle was a certain alternative to petrol engine vehicles, certainly for heavier transport. The diesel engine was also viewed as more cost-effective in the United Kingdom against the trolley bus. Its 30-to-35-year life cycle coincidentally came to the point of refurbishment and replacement in the mid-1960s, having been established during the 1930s. The vehicles required change, and the catenaries had reached the end of their life cycle.

Even so, the arguments were complex and, as we have learnt over time, lacking in crucial evidence regarding the environmental cost-benefit, which, in the long term of one against the other, has its penalties.

A report by The Electric Tbus Group(10) sets out to compare the main attributes of electric bus operations in cost-benefit terms for operating electricity-orientated transport systems, using Vancouver and Toronto Transit public service vehicle data and experience as an example. The two authorities, Vancouver and Toronto had a dilemma in deciding which of three vehicle type options would best serve their respective city's requirements - BEB (A) , BEB (B) or In Motion Charging (IMC) [the trolleybus]

What is a reasonable basis for comparison regarding electric-powered transportation adoption in the 2020s? The overriding consideration seems to be to avoid prejudging the end product of the analysis.

The most popular solution could be chosen because it puts vehicles on the road that are already in production and, on the face of it, appears to require very little infrastructure to support them. The Battery Electric Bus (BEB) [to use the report's definition] is, at first glance, a simple solution that has its attractions in its perceived simplicity. The central precept is that the battery electric bus requires no obvious infrastructure such as overhead wires, which is probably the main reason why many engaged in comparison immediately reject the trolleybus principle out of hand. However, is it as simple as that?

The illusion of simplicity is compounded by the fact that historically, in the late 20th-century application of the appropriate science, the trolleybus could not operate without the overhead power connection (a catenary). In the 21st-century application of the principle, modern trolleybuses can operate without overhead wires as battery-buses. Traction motor power electronics give full control of speed and braking, batteries, and onboard charging facilities. They are all shared, and in common, between all BEB, battery supplied electric-operated vehicles in all three guises. Their footprint on the thoroughfare is the same, and they carry no weight advantage one way or the other.

The power consumption of each type of vehicle is very similar. However, the difference in how the power is stored and transferred from its source to that onboard stored source (the batteries) and then into the drive mechanism of the vehicle is quite varied. The mode of power transfer and its onboard utility presents a fundamental element that needs to be considered when considering the capital investment required in operating those three varying types of electrically powered vehicles and their own technical variations.

The stated aim of the Tbus report (10) in comparing Toronto and Vancouver, the topography of the two locations, the distances the target vehicles need to travel coupled with the respective operational flexibility of the vehicles involved form the core criteria needed to establish a comparator. Those that are commissioned to install a system of vehicles and infrastructure that will run for 1,000,000 miles in-service for 25 to 35 years and provide the best value for money over its lifetime have all the relevant facts.

The trolleybus of yesteryear is a generation away from the trolleybus of the future, which is more accurately described as an 'In Motion Charging' (IMC) electric vehicle. In other words, the vehicle has independence from the overhead catenary when it journeys 'off piste' and is powered by battery alone. Therefore, the controversial element lies in how the battery system is charged in use and the effects of the charging facility in all relevant dimensions.

The Tbus report specifies three fundamental mechanisms currently available for running electrically powered buses, which are described as follows:

A. A BEB system based on slow overnight charging.

B. A BEB system based upon fast en-route charging - often called 'opportunity charging.'

C. A battery-trolleybus system using partial route overhead wiring and 'In Motion Charging.'

The purpose of the report is stated: "to cause transit planners and politicians to avoid automatic assumptions on the question of bus electrification." If it does do as it states, it will have achieved its purpose. (10)

Due to the divergence of circumstances in local areas, there is no conclusive proof that one mode of operation is better than another. Therefore, the local planners are urged to use objectivity and the insight provided by the report to come to a relevant conclusion. There are elements of commercial confidentiality that the report cites as reason for an inability to compare costs one with another for specific assessment. However, the paper has access to general costs, which do lend themselves to the required comparative elements.

The report costings are all in Canadian dollars; however, that makes little difference to the UK market when observing the comparison criteria. This is a report that is comparing two large-scale, long-term applications of the comparative criteria being used. Scale is all-important in every facet of the argument that is considered. Both cities aspire to purchase a significant fleet of vehicles in their respective locations achieving the best value for money from the public purse and confidence of certainty in operation.

In that regard, therefore, the argument will be very similar to any competing cities in the United Kingdom endeavouring to provide environmentally sound public service vehicles over the longest period of time practicable, with confidence in the continuity of service.

<u>Bus electrification options</u>

The report seeks to examine real-life examples from historical evidence. In this respect, the comparisons are uniform because both Vancouver and Toronto are substantial cities of approximately the same size.

The report initially considers the Toronto Transit Commission (TTC) proposal for a garage for 300 BEBs using overnight charging. [TTC commission green bus technology plan update.pdf](22)

The bus operator for Greater Vancouver Area, Translink, report on bus electrification until 2030 provides a complimentary comparison view. [Translink agenda mayors Council public mtg.pdf](23)

The comparisons used in the report used three options for examination:

<u>Option one</u> is a fleet of 300 BEBs using depot charging as proposed by the TTC.

<u>Option two</u> is for a slightly smaller fleet of BEBs using en-route charging.

Option three is for a fleet of battery-trolley buses with IMC sufficient to cover the same workload as options one and two.

Costings for a trolley bus system using battery-trolley buses with IMC using a streamlined overhead catenary system were obtained from a number of cities and detailed in the appendix to the report. (10)

Experienced consultants generally recommend that BEB fleets of vehicles using depot charging will need to be at least 20% larger than a diesel or trolley bus comparison. In this instance, with the report focusing on 300 BEB depot-charge vehicles, a perfect numerical ratio is obtained using 250 vehicles in respect of option three as a direct comparison to those 300 vehicles used in option one. It is worth noting that the Vancouver trolleybus system currently operating has 262 trolleybuses which the report puts forward as being close enough to establish real-life cost comparisons.

Option two, if it were adopted, is said to be capable of operating the same workload as option three with a 12% larger fleet than option three, which the report concludes would be 280 BEB route charging vehicles.

Suffice it to say that the report (10) then proceeds to illustrate in some depth the comparison of costs between the three options outlined above in respect of a variety of inherent costs. It is an extremely interesting document. The comparators certainly relate to what could be expected to pertain given a similar situation in two large cities comparing notes in the United Kingdom. Martin Wright, the author of the report, concludes that the figures do not prove that one type of vehicle is more economical in service or better in any sense than another because, he states, situations do vary from city to city. However, the analysis outcome from the report's comparative elements concerning the three possible approaches to large-scale bus electrification produce costings are outlined below:

Option one: battery bus fleet depot charged 300 vehicles - total capital cost C$446 million.

Option two: battery bus fleet en route opportunity charging 280 vehicles - total cost C$451 million.

Option three: trolleybus fleet In Motion Charging 250 vehicles - total cost C$405 million.

Given that all three options were of a similar magnitude regarding order size and implementation together with infrastructure capability, the author notes that In Motion Charging vehicles should not be discounted. The evidence is such that the often-ignored 'option three' BEB can be financially competitive with the other two alternatively charged Battery Electric Buses. Therefore, circumstances need to be taken into account in order to extract the positive nature of one application, one option, against another. Vancouver is remarkably hilly. Toronto is a relatively level environment.

Perhaps the availability of auto-connect and disconnect for the IMC-trolleybus-style application will bring the argument for considering that form of battery-powered bus even further to the fore. The viability of one form of vehicle propulsion against another takes on a distinctly different approach than pure financial costing. The power requirement, one environment in comparison to another, is fundamental to the final decision.

Chapter 14.

The *Bus Back Better* Initiative

Boris Johnson's opening comments in the *Bus Back Better* paper foreword on page 4 of the document stating that "he loves buses" are in no doubt. The transformation of Transport for London (TfL) in the early days of the *Routemaster* innovation certainly proves that point. Money was spent, modern technology employed, and passenger travelling became a joy until the reality of that bus's operation in-service data proved that there was still some way to go to achieve the radical change needed in the public service vehicle sector's overall efficiency in service.

The foreword goes on to state that usage of public transport in London has grown by more than half. (page 4) Yet, in a subsequent analysis,(page 9) it states, "the (sic London) bus ridership has been falling." A contradiction in terms maybe, but nevertheless a statement of fact based on nine years in service for the New Routemaster vehicle and its successes - and failings. However, the New Routemaster is not the only bus in the TfL fleet or other operator in London, where there are 9,300 buses, over half of them diesel-powered.

£3 billion is to be made available, and 4,000 new 'green' buses are to be introduced to the United Kingdom bus fleet. However, does that beg the question, which is not, "how quickly can you spend £3 billion?"

Rather, the burning question is: how many new buses can you manufacture in any given 12-month period? How long does it take to build a new bus from scratch? Given the global interest in new buses, particularly in Europe and the Americas, how available are the rare minerals required for the batteries that are necessary to power those buses? Ford Motor Company have had their production line of battery-driven cars brought to a standstill because of a hold-up in specialist electronic components (Spring 2021).

The opening pages of the *Bus Back Better* paper(page 8) quite rightly highlight the significant status of the double-decker bus in Great Britain and Northern Ireland.

Buses are the quickest, easiest and cheapest way to reduce carbon emissions and improve their quality,(page 8) but there is a timescale that goes alongside the government aspiration to provide clean air. The New Routemaster vehicles for TfL took years to manufacture and deliver 1,000 buses. The 4,000 bus aspiration will certainly not be achieved universally without the continued development and production of Euro VI diesel buses, preferably in a hybrid format, for many years to come, if services are to be maintained.

Indeed, rural country routes can accommodate the impressively reduced emissions from Euro VI diesel engines at some 6% of their predecessor's emission tally. In contrast, urban areas would still find it difficult to accommodate any emissions willingly due to the number of vehicles involved. If that number of 'unclean' vehicles could be reduced, it would be a significant starting point. Therefore, why not accommodate the heavily populated areas and the less densely populated suburban and rural areas concurrently with the same vehicles operating efficiently in different modes of propulsive power?

There is a balance to be achieved and, at the moment, with over 15,000 diesel buses that do not meet the Euro VI standard (8) on Great Britain's roads, a significant diversification of capital spend from that promised £3 billion needs to be achieved. The manufacturing capacity gap could be bridged by spending on higher specification diesel engine replacement hybrid vehicles and upgraded hybrid diesel vehicles whilst concurrently reducing overall emissions in a planned manner.

The Bus Services Act is specifically quoted, (page 12) but no reference is made in the *Bus Back Better* document to the planning permission legislation that halted the Leeds 2016 initiative to transform that city's public transport system. That legal impediment would adversely affect far simpler and more cost-effective schemes than that proposed for the Leeds transport initiative. This will cost time, effort, and money unnecessarily unless it is altered to permit, say, a simple catenary system infrastructure to be placed within a city or urban environment to provide 'in motion bus battery charging' from overhead facilities - if that solution was deemed most beneficial.

The battery bus fleet can benefit from a catenary installation, but current planning laws will thwart the best intentions of local authorities wishing to take advantage of such a system being implemented for either the hybrid or battery fleet. There is an urgent need for those constraining acts of parliament to be withdrawn and/or modified within that £3 billion budget if the money is to be spent effectively and the timeline stated is to be met.

The appropriate planning legislation will allow the 2017 Bus Services Act to be brought into play efficiently by the Traffic Commissioners. Therefore, the notion of Enhanced Partnerships as outlined in the *Bus Back Better* paper (2) can certainly work to great advantage in improving passenger experience and hopefully bus service integration with other forms of transport, i.e. rail and tram services, as appropriate.

Priority given to schemes linked with an overhead catenary system would achieve faster, more reliable and cheaper to run bus services through urban and city environments than the current fleet can achieve. Funding is a delicate balance between new vehicle acquisition and fleet allocation, coupled with upgrading and modification by retrofitting older but still serviceable vehicles which can be refurbished during the retrofitting process. Those vehicles can then be brought back into service with lower emissions (suburban/rural routes) and zero emissions (urban and city routes) using the overhead 'In Motion Charging' (IMC) system for onboard battery charging and propulsion. The Euro VI diesel engine can still be employed outside of the central city or urban part of the journey, giving vehicles a much more useful versatility as well as a continuous passenger experience to major conurbation destinations.

The linking of suburban starting points with urban destinations such as railway stations, bus terminuses, university sites or technical colleges, hospitals, schools, workplaces and so on by way of an onward 'through urban' back to suburban destinations rather than making the city centre the ultimate hub/destination for all buses coming in from all directions, would greatly improve the passenger experience. When linked to a catenary system using retrofit buses, it results in a vastly improved emission-free city centre transit area for those vehicles to pass through; rather than to.

The electric bus has yet to prove that it is cheaper to operate on a battery basis than its diesel alternative. The statement (page 13) in respect of the 4,000 new 'zero-emission' buses would be less controversial if those new buses were capable of recharging their batteries whilst conducting the PSV role of moving passengers from A to B. The reality is that they are stationary for even as little as 6 to 10 minutes, charging their batteries from overhead gantries that are clumsy and expensive to install.

There is a need for overnight charging using far more sophisticated, expensive, and, some would argue, time-consuming infrastructures compared to the simplicity of a catenary that could be provided by the local authority. In Motion Charging is then achieved trolleybus-style by means of a retrofitted unit on the roof of each vehicle to facilitate the application of power to a 'second life' bus.

TfL found that the cost-benefit of all-electric vehicles was not as originally thought. There are so many facets to the comparisons of cost against benefits equation to consider. It transpired that providing the power charging facilities TfL needed to keep those buses in service was more complex than anticipated. Originally Shepherd's Bush garage was the assumed base for the electric bus fleet; however, the local provision of electrical power proved unable to accomplish the desired outcome. A special, purpose-built facility was required.

Comparison of in-service data and the initial cost of electric vehicles against Euro VI diesel engine vehicles and a standard hybrid counterpart really makes its mark over time in fuel savings. The impediment to that cost-benefit analysis is in the consumption of batteries. Particularly, the foreshortened life of those batteries brought about by high-density charging as opposed to trickle charging, which can be achieved through the retrofit initiative and an overhead catenary system. The 2013 EU report illustrates a substantial cost for new batteries every three years for battery-powered buses.(1, table 4.1 p 67)

The table illustrates the cost for replacement batteries and associated equipment for every 10,000 charges. It is estimated at three years replacement frequency that a cost of €150,000 - €200,000 per vehicle over a 15-year period is expected compared to half that for a catenary-connected vehicle.

A catenary is designed to provide In Motion Charging, and the Department for Transport analysis shows an average cost ratio benefit of 4.2% of the cost. Just think how much more value for money can be achieved with the catenary and an increased battery life of 50%. Instead of seven battery pack changes (purchases plus fitting time out of service), there would be three such requirements: a significant cost saving over the life of the vehicle.

The target date set by the *Bus Back Better* paper is the end of June 2021. By this date, it was expected that all Local Transport Authorities (LTAs) except municipal authorities woud have started the statutory process of franchising bus services if appropriate. They will at least have committed themselves to establish Enhanced Partnerships across the authority's area under the Bus Services Act throughout the process. It is not impossible for local authorities to pursue franchising, and they may choose to do so.

However, the paper does point out that the authorities have to commit to implementing Enhanced Partnerships in the meantime until that franchising process is complete. *The Failure of the Rail Franchise* (19) proves to be helpful reading.

The process is therefore already underway, and the expected delivery of Enhance Partnerships is April

2022.(page 14). Are the timescales realistic? Some local authorities with larger or more aggressive legal departments could outshine smaller or less intuitive local authorities. Doing so undermines the 'levelling up' process by affectively excluding those slower pace authorities from the £3 billion grant fund. The COVID effect has also slowed everything down. These dates may simply be too ambitious in the circumstances for the desired effect of achieving the twin outcomes of 'net zero and levelling up' as outlined.

Bus use over the period 1985 to 2019 shows a marked decline in journeys, but the graphic shows little information regarding passenger usage on that declining journey profile. (page 21) Approximately half of the Great Britain bus fleet (and 100% of the Welsh bus fleet) are pre-Euro VI diesel engine vehicles. That indicates that in excess of 15,000 buses need replacing urgently. Not only are they older (is that why the public are not using them?), but they do not meet the latest Euro VI emission standard some 13 years after it was introduced as an aim for a minimum in-service specification.

Much of that 'below standard' bus fleet is also perhaps in need of upgrading internally. An assessment is urgently required, by engineers rather than accountants or administrators, to establish whether those vehicles that are below the Euro VI standard can be upgraded and given another ten or more years of life. Perhaps this could be done by a modification to that higher engine specification standard, or maybe a combined hybrid standard capable of being in motion charged (IMC) as outlined in previous chapters. What is the point of taking buses out of service in an area well catered for in environmental terms and replacing them with electric vehicles if those same buses exacerbate the problem in another area of the world in the second-hand market?

Surely a program of upgrading and rejuvenating all the vehicles where practicable, retrofitting them to utilise zero-emission power for as much of their journey time as possible and then reintroducing them to through city routes to optimise their passenger experience would be a more practical proposition and better value for money than relying solely on newly manufactured vehicles?

That £3 billion is not going to go very far if it is all spent on new vehicles (which are going to need new batteries in 3-4 years' time). The proposition requires a cost-benefit analysis such as the Canadian-centred report from Tbus (10) to consider the whole fleet such that 'net zero and levelling up' can be achieved in an optimum time delivering value for money to the most out of date areas first.

A further consideration is in relation to the purchase of new vehicles. Should they all be 100% electric vehicles? Electric vehicles cost 150% of the price of fossil fuel vehicles, whether they are hybrid or electrically driven. A glance around any car or light vehicle showroom indicates that the purchase of one's own mode of transport, and perhaps the two-car family, is on the wane.

The quicker the improvement to public service vehicle provision takes place, and the more acceptable those vehicles are to potential passengers; the less likely individuals are to take on the financial commitment of a personal vehicle purchase. One only has to view a contemporary housing estate to realise that parking facilities for cars are becoming more and more difficult. Even people whose houses have garages do not use those garages to store their cars. The garages are storage areas to augment the modern, reduced living space property.

Consider a road of parked vehicles with electric cables attaching to a large number of those cars and vans overnight with no garage protection. It presents an open invitation to metal thieves for precious metal cables. If those thieves perceive that stealing from railway embankments is fair game, they will steal from anywhere that quantum justifies the risk. NOTE: Network Rail replace copper signalling wire every few days from parts of their network. Yes, days. the frequency is determined by thieves stealing the trackside copper signalling cable. The purchase of electric cars is far from assured with that attendant future risk profile.

The fewer cars there are on the roads, the more public transport will be in demand. The public will decide.

Surely the **target** over the next 20 years to 2040 has to be the **significant reduction** in toxic effluence rather than its elimination. Elimination is an aspiration that simply cannot take place with the resources available and the number of vehicles involved in the time scale aspired to. A major new industry is required within Great Britain and Northern Ireland to make these dreams, for that is what they are, come true. The retrofitting and refurbishment of existing vehicles is an answer that can help meet the target set in this ambitious paper, *Bus Back Better.*(5)

The In Motion Charging (IMC) and propulsion proposal can operate in a number of ways to suit the local environment and the aspirations of the local authority involved. The most effective means of charging the greatest number of buses in service over the most efficient route is achieved by installing a catenary system through the city or urban centre. This allows the largest number of vehicles to use the system to recharge batteries whilst also providing zero-emission, full-power propulsion, for the transient vehicles. This is particularly pertinent where the topography is such that battery-powered vehicles would have difficulty in negotiating inclines to either enter or exit the urban area.

The alternative would be for a more extensive but more easily installed catenary system on the entrance roads to the urban area (i.e. in suburbia or the countryside). It would allow for battery top-up prior to full battery use in the city centre and on the first stage of its journey out of the centre. There it would re-engage with the catenary to recharge the batteries prior to disconnecting and travelling further into a rural or a more heavily populated suburban destination such as a housing estate or industrial estate/major employer factory processing complex. The third option would be a marriage of the two principles for which the author can see little advantage or rationale in utilitarian or cost-effective terms.

The biggest problem underpinning the ambitious nature of the United Kingdom's Prime Minister's initiative is seated in the ability of British industry to produce sufficient vehicles to meet the ambitious targets set by the paper.

One only has to look objectively at the number of buses that can be produced as new vehicles in any 12-month period. If that figure is coupled with the number of existing fleet buses that could reasonably be taken out of service and upgraded to meet the lower emission standards required on the net-zero initiative, the time taken to meet the target on current production figures will approach ten years at least. Couple that vehicle production element with the infrastructure changes and planning processes required to install those vehicles commensurate with the 'levelling up' process, which is central to this paper, and one starts to comprehend the difficulties. Can British companies rise to the challenge of those two initiatives?

In terms of bus production in the UK, as evidenced by new registrations of buses, (20) production at its peak from Wrightbus, a major manufacturer, was just over 230 per quarter at its peak according to the 2020 DfT report on vehicles licenced.(20)

A further question relates to the most appropriate way to upgrade the existing half of the bus fleet below the Euro VI standard to meet the objectives outlined in the paper. Surely if the facilities available can produce seven or eight buses at the same time/ cost of one new bus and in so doing upgrade the existing fleet rather than disowning it or discarding it (selling it somewhere else which merely moves the problem to another part of the globe), that must be worthy of consideration.

The visual impact of a catenary pales into insignificance when one considers the results of the 2013 European report,(1) and especially the results of the in motion charging analysis which examined the major means of operating electrical buses in two eminently comparable circumstances involving major Canadian cities. (5). The conclusions drawn from comparing a bus fleet using depot charging and requiring 300 vehicles to achieve the operator's route objectives with a bus fleet using en-route opportunity charging stated as requiring fewer vehicles (280), both of which had a total cost comprising infrastructure and vehicle of circa C$450 million, must carry some credence.

By further comparing those attributes and costs to an In Motion Charging program, which reduced the number of vehicles required (because there was less time out for charging) but, of course, required a catenary to provide the

In Motion Charging structure for the vehicles to use, at a total cost of circa C$400 presenting a saving of C$50 million, has to result in the serious consideration of all three modes of operation.(10, page 6)

It is pertinent to quote Martin Wright, the author,(10) when he states, "as all three options are of a similar order of magnitude, it is important that transit planners and, in particular, political decision-makers be aware that battery electric buses are by no means the cheap and simple solution that many seem to think. Conversely, and contrary to appearances, an installation using battery trolleybuses utilising In Motion Charging (IMC) can be a financially competitive option using those battery electric buses."(10)

The principal answer to where the money should be spent does not lie in the provision of bus lanes or the purchase of new vehicles per se. The primary objective must surely be to put in place a system that will provide a service that the prospective passengers will use, and the local residents will benefit from environmentally. The decision should not be determined by appearances. The installation criteria should have its initial focus on major through-routes, linking all points of major conurbations and then connecting those population centres with the railways, hospitals, schools, and workplaces, with where people seek service, work and shop amenities.

If equipped with a catenary system, the installation on at least part of the routes will allow hybrid vehicles to enter and work in the urban city centre with zero emissions on full-electric power whilst getting their batteries charged as appropriate for the free-running battery or Euro VI diesel-powered journeys. The addition of Euro VI diesel engine facilities is a necessary bonus that should not be ignored. Higher journey speeds, coupled with reduced stopping and starting, can surely make the minimal emissions the facility can achieve worth accepting until a fully-electric fleet can be established over time, if appropriate. (5, page 30)

Chapter 15.

An Example of What Is Possible - Lincoln

Lincolnshire County Council created the InterConnect and CallConnect services in Lincolnshire, formulating the concept of rural passengers being transported by minibus vehicles to meet the local main bus service in line with their published timetable in 1997. By 2013, five 'InterConnect' bus routes had been established to join the major towns surrounding the industrial City of Lincoln and accommodate the surrounding rural agricultural and holiday environments. A CallConnect minibus service was integrated to bring passengers in from rural locations to join the main bus routes. The InterConnect philosophy/concept is a core element of the March 2021 *Bus Back Better* DfT paper, which is reassuring.(5)

The original route services were numbered 1,3, 5, 6, and 100. (Illustrative Map 15.1) Those services, some entailing journey lengths of well over an hour to and from Lincoln, were operated under the Lincolnshire Road Car banner, which inaugurated the bus element of the scheme. The Lincolnshire Road Car company was sold and eventually became part of Stagecoach East Midlands, which now has its headquarters in Lincoln. The 100-route bus service stayed as it was numerically and continues to provide the north west service from Lincoln City through Saxilby to Gainsborough and on to Scunthorpe. Route number 1 also retains its numerical identity and continues providing its service along the A607 from Lincoln City south to Grantham, passing through the 'Trent Cliff Edge' villages.

The route number 3 service became route 53, Lincoln City north east to Cleethorpes via Market Rasen. The route number 5 service became route 55 and joined Lincoln City southeast to Boston via Woodall Spa, and finally, the route number 6 service became route 56: Lincoln going east through Horncastle to Skegness.

It is those five entities and the long-established InterConnect service in particular that are being considered here. All of the services are being operated. However, Stagecoach East Midlands in 2021 using diesel buses with registration plates containing 07, indicating that they were first registered between March 2007 and August of that year. The current Lincolnshire fleet does not contain any hybrid buses. For all statistical comments, please note that a bus is defined by the Department for Transport (DfT) in their statistical parlance as a double-decker bus. A single-decker bus is a 'coach.'

'Minibus' is the third category of vehicles of up to 18 seats.

The following points arise from one or two assumptions, the first of which is that clean air is essential in central Lincoln. The centre of the downhill commercial centre is where all of the current diesel buses congregate at the bus station, which is also their terminus. There will be many other large towns and smaller cities that fit the Lincoln profile. Therefore Lincoln is a good example of an ageing bus fleet and a core environmental hazard that is reflected around the United Kingdom. The assumption here is that the InterConnect buses (DfT defined)(5) are replaced by hybrid buses with a battery duration that would carry the vehicles through their route journey's major towns and extended villages on battery power.

The second assumption is that the buses would travel through Lincoln city using 'In Motion Charging' (IMC) for battery top up and propulsion for the vehicles. A further assumption is that those vehicles would comprise a through city operation rather than terminating in the current terminus bus station.

A catenary system would be introduced to the city centre in order to entirely eliminate diesel emissions from the vehicles operating on the InterConnect routes. That would effectively extend the current facility for passengers to travel to Lincoln and continue their uninterrupted journey through Lincoln to destinations the other side of the city, should they desire. Single vehicle destinations may include the university, the technical college, secondary schools, medical facilities, hospitals, the central railway station, H.M. Prison and the major employment and residential estate locations that people may wish to travel to on the city outskirts.

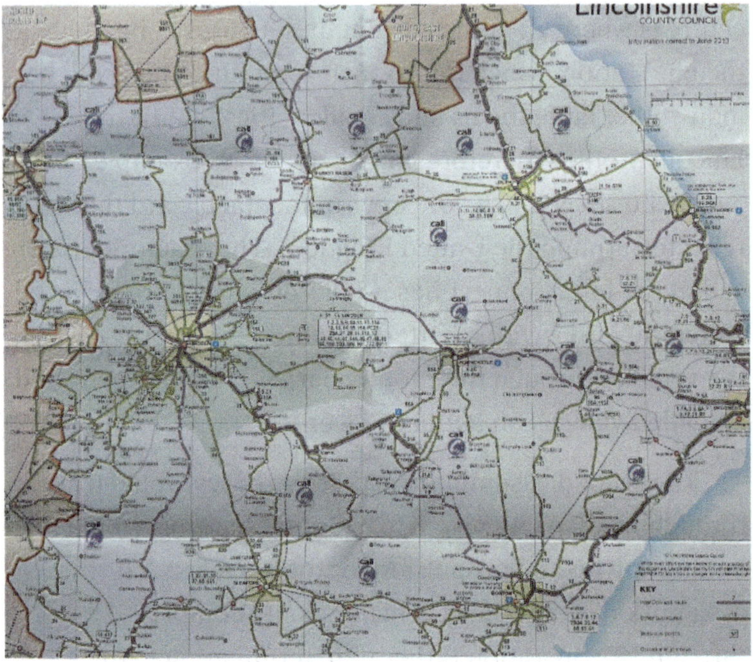

***15.1** Illustrative Map - Lincolnshire County Council – extract from bus timetable 2013.*

Those through-services could be varied at the central railway station adjoining the bus station and Lincoln central High Street as a connection point. A change of crew would then take the vehicle on to its further destination and, of course, renumber the bus according to the ongoing route. In this way, the number 1 service could start in Grantham, proceed through the A607 villages into Lincoln and connect to the catenary at, say, Waddington. The service would continue through Bracebridge Heath and down past the secondary school on Cross O'Cliff Hill into the city centre via the High Street, Tentercroft Street, and on to the central station.

The bus would stop, disconnect from the through-catenary line and immediately reconnect to the short catenary feed positioned at the bus stop. (See diagram 15.2.) At this point, the bus would renumber to say route 56 and proceed to exit the bus stop feed and, via a simple pressure point connection, re-join the main catenary.

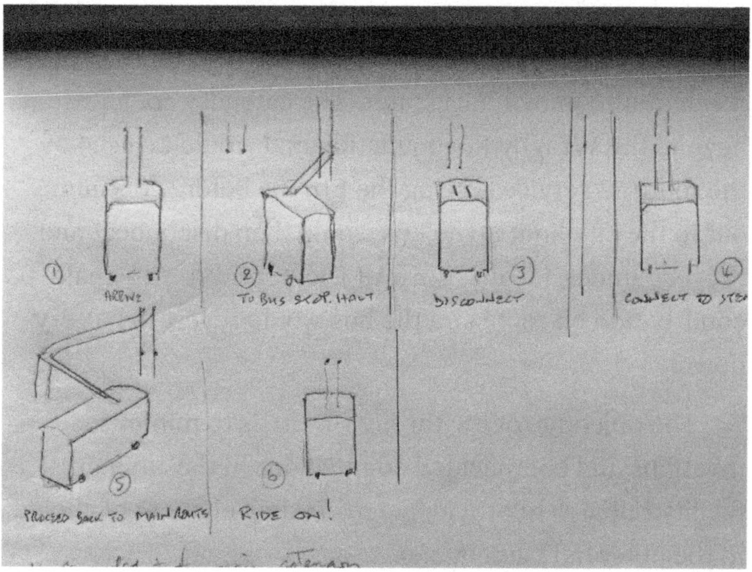

***15.2** Illustration of the reconnection facility allowing passing trolleybus traffic.*

1. Connected to the main catenary.
2. 2 Still connected, enter the bus stop layby.
3. Auto disconnect from the main catenary.

4. Immediately redeploy poles to connect to bus stop catenary feed.
5. Bus moves out to re-enter the main catenary.
6. Drive on under In Motion Charging (IMC) power.

Following a crew change, the route 56 designated vehicle could then carry the passengers via Silver Street (unconnected) at the north end of the High Street. Then, when reconnected, continue up Lindum Hill under full 'in motion' catenary power, past or near to two senior schools, the hospital, H.M. Prison, and the outer circle residential/commercial estates. The catenary could extend beyond the Wragby Road junction and would be used by the route 53 service serving the Ermine before continuing out to the city limits. The bus would then disconnect and proceed under, ideally, Euro VI Diesel power. Any major conurbation en route and the bus would revert to battery power.

The objective of the through-route is to minimise junctions and complicated connections in the catenary facility but also to provide an overriding element of choice to the potential passengers.

It is great that there are advertising initiatives on television or local radio networks indicating the efficiency and scope of the bus network nationally and locally. This feature would notify the public of the upgrade to potential advantage once the revised operational methodology has been established.

The route 1 to route 56 bus services could perhaps alternate between a route 1 to route 100 bus service with that second bus travelling out to Gainsborough and through to Scunthorpe. The current services are hourly, and the alternate service could be operated as postulated here to give passengers improved destination choice. Similarly, the return journey would carry passengers seamlessly to the central station, university, etc., without the need to transfer from one vehicle to another. Many medical patients require to visit diverse hospital facilities in Boston, Grantham, Gainsborough, Scunthorpe and Lincoln, plus minor centres in small town locations, all of which are currently served by these five bus routes.

The essential asset attributed to buses connected to 'In Motion Charging' (IMC) is that the electrical 'fuel' is less expensive and has the 24-hour capacity to power the bus system. Therefore, the working duration of the bus is extended, and the in-service capacity/earning power of the vehicle is optimised. A battery-powered bus may serve an 8-hour shift and then require 3-4 hours of charging, dependent upon the facilities utilised.

Full battery-powered buses would have difficulty negotiating the Lindum Hill and Cross O'Cliff Hill inclines due to the power requirement from the batteries being less than adequate to cope with the gradient. The vehicles would require hasty and time-delaying battery top up charges along the journey - if that methodology was adopted. The provision of route-installed chargers necessitates a significant infrastructure cost (10) and incorporates substantial, if short, time delays to the journey.

High-speed charging is proven to be detrimental to the life and overall duration of onboard batteries, thus increasing operational costs and fleet size unnecessarily. That impediment creates a capital cost every 3 to 4 years for new batteries, whereas using In Motion Charging reportedly extends battery life to 5 to 6 years. That represents a significant cost saving as well as journey time benefit, as evidenced in the Canadian paper (10) discussed in Chapter 13. The EU strategy paper from 2013 makes the same observation in its analysis. (1, page 69)

The foregoing suggestions provide potential passengers with a means of getting from villages south of the city to family, friends or work located in the north or east of the city on perhaps a two-hourly through city service, rather than having to change. This would not deny them the hourly facility if a change in the city centre is appropriate, as is available now. The main advantage, however, is cost and utility.

By presenting a through-city transit facility (it is up to the operators whether they take up the idea or maintain the status quo), buses using the IMC would not need a turning circle in the city centre; they would merely continue a journey or complete a circuit. Those travelling in the opposite direction would similarly benefit.

By way of a note, the central bus station terminus philosophy currently adopted by many large towns and cities does provide a means of public engagement and convenience for the traveller. However, having a central terminus, unfortunately, creates a lot of the current emissions that are the subject of the present dilemma.

The proposed 'through-route' changeover should at least be considered. If the catenary system and the St Mary's Street through-route is adopted, then vehicle emissions will be reduced dramatically in the city centre. Further benefits will be felt en route, courtesy of the catenary's introduction and improved battery performance statistics. The revised location - adjacent to the current bus station and between it and the railway station - provides an ideal confluence of routes and transport styles as promulgated in the *Bus Back Better* DfT paper. (5)

Eventually, the system could be extended to bring in other route permutations from the Birchwood area, west of the city, and perhaps the new Swanpool development. This could be by way of the service route 15/16 service alternately coupling with the route 11/12 service or route 53 service to Market Rasen/Cleethorpes, which would follow the same procedure at the central station catenary 'feed' facility to be utilised.

Passengers would have a service to and from the Birchwood highly populous area of Lincoln to the hospital and working/residential conurbations in the north, through to Cleethorpes alternating with the

Grantham/Gainsborough Scunthorpe service. The overhead In Motion Charging catenary would extend from Birchwood/Swanpool to the west of the railway track, joining into Newland. Another prospective route could connect to North Hykeham railway station and stay on the eastern side of the track on Triton Road over the University Bridge to Newland and through to the central railway station. This would provide connecting services, again, as outlined in the 2021 *Bus Back Better* paper from the DfT. (5)

The catenary route would be a simple, minimum junction assembly with possible crossover junctions at points where, on a rare occasion, two routes might require to cross each other to provide continuity of In Motion Charging for the bus routes involved. An estimated cost of £1.2 to £1.4 million per mile of infrastructure implementation, coupled with the estimated £42,000 per bus retrofit and refurbishment costs in the Lincoln environment as outlined in Chapter 16, (Tables 16.1-4) would represent excellent value.

A scheme incorporating the acquisition and retrofitting of, say, 24 vehicles would involve a modest outlay of £1.008 million added to, say, 8.5 miles of infrastructure at between £10.2 and £11.9 million, which is surely worthy of consideration. That system could feasibly be in place and operational in a shorter period of time, planning permissions permitting, than manufacturing new and expensive vehicles coupled with any infrastructure costs attendant to the system chosen.

The capital cost of 30 new all-electric buses (option 1 or 2) Battery Electric Buses (BEB) would be £13.5million (at £450,000 per vehicle) plus their attendant infrastructure costs of £10.7 million by comparison.(10)

The timetabling would enable the optimum use of the 24 vehicles perceived to provide the connecting facilities with virtually all of the major industrial and domestic conurbations with a minimum hourly service, or more frequently if the revised time scheduling permits on appropriate routes. Electrical power would be used in the high population density areas, thus reducing the environmental impact to zero for a substantial element of the bus fleet and thereby cleaner air for the population on route. Any or all of the five route services could, of course, be upgraded to a more frequent service, alternating their commencement points and designated destinations.

Chapter 16.

Detail of the Proposed Scheme Layout - Lincoln (or Your City)

Having discussed and approved the general dynamics of the aspirations for an In Motion Charging facility, the facts now require assembly. This is the start of the project proper.

What do the planners want to achieve?

In what timescale do the planners want to achieve it.?

That surely is the opening gambit. In the case of the Lincoln example in the previous chapter, there is no better place to start than there.

The fact that Lincoln City Council, Lincolnshire County Council, and central government all aspire to achieve the same thing for the city, county and country, which is clean air, coupled with an efficient bus service to provide for the needs of potential passengers, is encouraging. The replacement of the current sub-Euro VI diesel-engined bus fleet on heavily populated long-distance routes is a good place to start, taking care not to be over-ambitious.

Because of the nature and layout of Lincoln city and the surrounding countryside, it is an ideal vantage point from which to implement an assessment of the viability of a smaller fleet of public service transportation in a reasonably densely populated city environment. The city couples up to compact suburban housing estates and well-populated suburban and rural conurbations around its periphery. It is an ideal location to examine how the electrification of the bus network can also be achieved economically and environmentally in similar regional centres.

The broad outline in the previous chapter needs to be examined considering current regulation and legal boundaries to establish what impediments there might be to planning applications, in particular, the proposed route changes in general with local authorities, government, and, of course, the route operators. The responsibility for the funding and operation of the routes aspired to need to be considered carefully, particularly in light of the successes and failures inherent with the rail franchise system and, conversely, the nationalisation experience of previous times. Lessons are there to draw information from and act upon.

The initial scheme envisaged for Lincoln centres on upgrading the InterConnect/CallConnect system to make it as environmentally sound as possible in the 'short term,' with a view to an entirely zero-emissions system superseding the status quo in the medium to long term. The 'Rome wasn't built in a day' syndrome should be an acceptable premise at this time. There are just too many environmentally substandard buses operating nationally for the whole system to be converted to zero emissions by 2030. That assertion is based upon current manufacturing and upgrading facilities that are available in the United Kingdom in 2021. That being said, it is worth the effort to try.

The beauty of the aspirations for Lincoln and Lincolnshire is that they enable the most populous parts of the area to experience zero emissions from the buses passing through them whilst minimising any detrimental effect in the other areas on the proposed routes. This is achievable within a viable budget contained within a 10-to-15-year view to electrify the whole of those proposed routes in the most cost-effective way over the longer term. The infrastructure put in place now should facilitate the long-term (30-35 years) desired outcome.

How that can be achieved has already been outlined in part; however, costings and assumptions need to be defined specifically to be realistic and factual, with quotations being obtained to achieve the aspired results from known and new contractors/suppliers. What is proposed involves five known bus routes. That could expand to possibly a sixth or seventh in the midterm. The possible extra routes that InterConnect, in principle, with its current five routes, could therefore be a factor in common route vehicle upgrades over time. An example for consideration would be the 15/16 route servicing the town of North Hykeham south of Lincoln and / or the 11/12 services to Welton and Nettleham to the north.

We are best however containing our enthusiasm in order to accurately assess the initial phase. The five routes suggested concentrate upon the current InterConnect routes 1, 31, 55, 56 and 100. (You will hopefully appreciate the mention of the other two possibilities during the following analysis.)

The five InterConnect route applications have merely been outlined in the previous chapter. There is a need therefore to expand upon that outlined information here.

The initial phase would be to examine the regulatory and statutory obligations and restrictions that might apply and ensure that they are considered when consulting with infrastructure contractors and the bus operators. This must be done in conjunction with their suppliers and vehicle modifying agencies where the planned development of the services is concerned.

The Transportation Act 1985 could present a problem by way of the time taken to follow the stipulated processes. It is said to be an out-of-date piece of legislation presented to a forward-thinking local authority (as Leeds City Council found to their ultimate cost in 2016).

What will the infrastructure look like, and how complex will it be? That surely must be the first consideration. What is envisaged is a simple catenary system that provides In Motion Charging for a small but significant fleet of vehicles in use as they pass through the City of Lincoln. Whether the through-route philosophy outlined previously is adopted, or the consensus of bus operators and the local authority is that it is better and more efficient to operate the services on much the same basis as they have done for the last 20 years needs to be established. However, it should be done in light of a meaningful time/cost/benefit analysis coupled with the environmental factors, rather than a gut response rejection of any fundamental change which may actually be productive in a multiplicity of ways.

The current bus station activity is frenetic at times, with vehicles constantly moving through the day.

If the proposed electrically powered In Motion Charged buses pass between the bus station and the railway station linked to an overhead catenary, as has been suggested, then zero-emission bus transportation has a new and relevant focal point. The fact that the services operating also carry passengers to the northern end of the High Street as a destination, as opposed to the moribund route of outward-bound vehicles travelling north using Broadgate, is perhaps an opportune benefit.

Broadgate provides no boarding or alighting service of any meaningful relevance until the top of Lindum Hill. That, in passenger terms, is a factor that requires contemplation. The Mint Street/Silver Street highway would be battery-driven until the bus stop reconnect facilities are reached. The first 'feed' connection would occur beyond the traditional on street banner advertising feature and connect directly into the southbound route 1 and 31 services. The second 'feed' provides for the 53 and 56 north / eastern services to their respective catenary systems, as shown in diagram 16.1

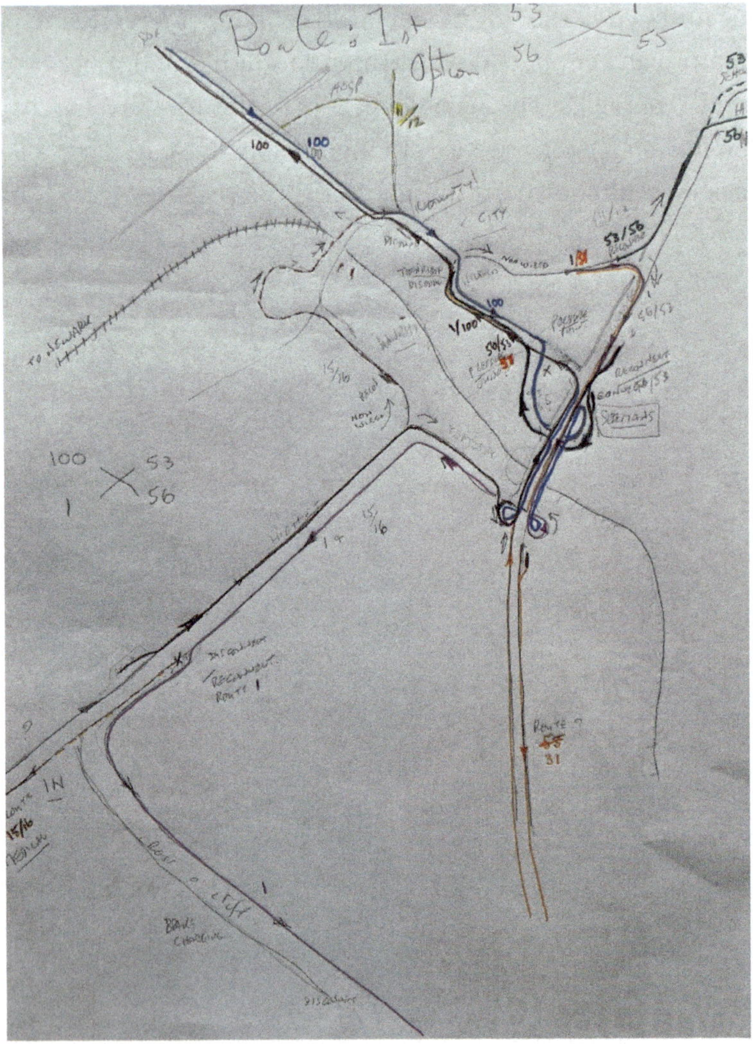

16.1 *The proposed catenary installations for Lincoln InterConnect bus services.*

Further out into the north and east, the main hospital, two large schools and H.M. Prison would all be on an environmentally sound bus route, as would the commercial areas and large residential areas on the north/eastern outskirts of the city.

Those bus route connections would be linear and also allow passenger carriage to the local authorities, both city and county, who would benefit greatly from a zero-emissions bus service providing for their employees from both the central station and the large village of Saxilby on the 100 service route. The buses travelling into the city from the south and northwest would facilitate public transport access to the northeast as outlined in the previous paragraph by way of a simple transfer of the vehicle to onward routes at the central station or the bus station, both of which would be easily accessed. (Diagram 16.2)

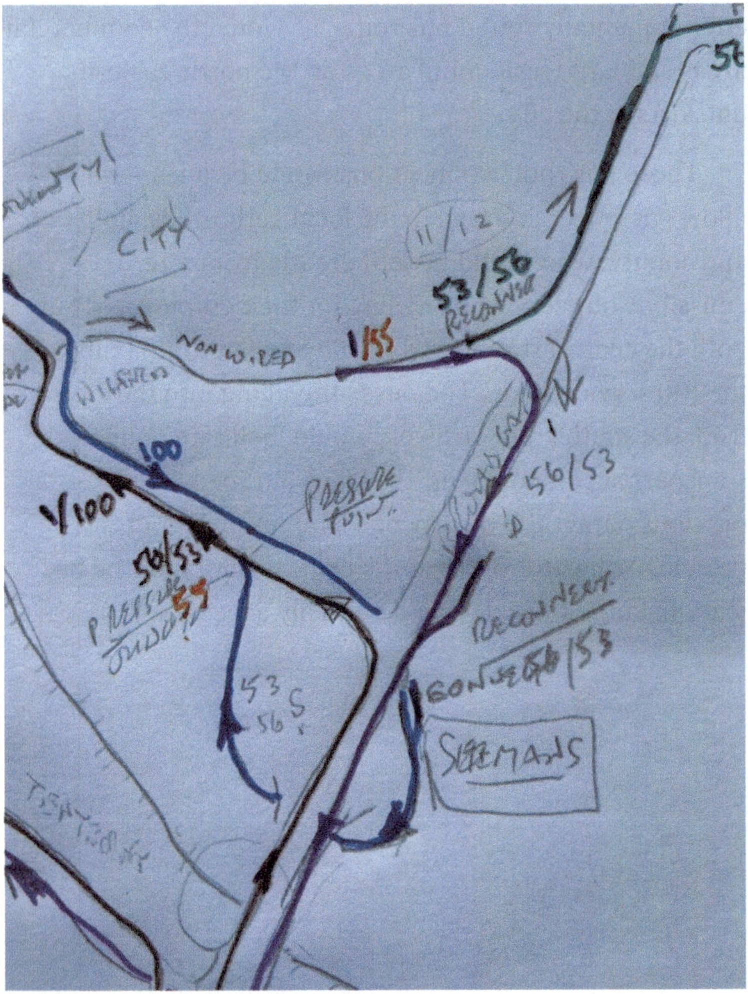

***16.2** A more detailed view of the catenary structure in the city centre [TPO'H]*

The through-routes from the north and the south connect directly to Siemen's factory, one of the largest employers, also the local authority central offices with all five interconnect services and the Lincoln Central railway network. The villages to the south and southwest, including the town of North Hykeham, would complete the connectivity circle by way of route 1 from Grantham and the 31 from Boston via Woodhall Spa, along with the already well-used North Hykeham rail connection, into and out of Lincoln central station.

Once the proposed routes have been decided upon, the next action will be to find experienced engineers and transport consultants to design the detailed infrastructure composition for both application and costing to take place. A visit to Sandtoft Transport Museum with its working trolleybus network would prove an enlightening and instructive experience.

Electrification of the bus routes appears to be a foregone conclusion from the most up-to-date reports. Therefore, the decision has to focus upon whether the dedicated battery bus options (BEBs) are the eventual vehicles to use or whether an Option 3: In Motion Charging trolleybus network would provide better long-term efficiency coupled with value for money.

Having outlined what the In Motion Charging (trolleybus style) network would look like, the number of lines and the number of buses required to provide a useful service, set in terms of the cost of the alternative service styles, needs to be assessed. The three battery electric bus (BEB) operating styles defined in the Electric Bus Group paper(10) are:

Option one: En route charging - charging points placed along the route at strategic points. We will call them Speed Charging Points (SCP).

Option two: Central overnight charging - a central Overnight Charging Facility (OCF).

Option three: In service battery charging on strategic elements of the route to slow charge whilst providing the service. In Motion Charging (IMC).

The analysis becomes central to whether an In Motion Charging (Option three) answer is applied in the Lincoln sample case. By way of a note, comparisons can be made with the cost of running a diesel fleet; however, that seems rather futile given the environmental targets that the authorities, are attempting to achieve.

Certainly, an environmentally reduced diesel element would be acceptable over, say, the next ten years whilst manufacturing capacity is increased and entire fleets of electric hybrid and/ or fully electric vehicles are produced, displacing 'compromise' vehicles (Euro VI Hybrid vehicles) to less-populated areas in a logical replacement programme to mitigate the environmental hazard proportionately over time.

After the transition period, exactly as happened with the historic equivalents (horse to motorised hybrid to petrol to tram to trolleybus to diesel and the present transformation), one would hope for a full, electrically-powered bus service on most if not all major bus and coach

(DfT definitions) routes, where they were able to deliver service. That currently involves circa 40,000 vehicles currently in service (5,6,7,8,9).

That would include those InterConnect services specified here, where at the end of the initial 10-year period, the infrastructure put in place for, say, Euro VI diesel hybrid buses retrofitted to In Motion Battery charging would be replaced by a dedicated new fleet of In Motion Chargeable battery electric vehicles as manufacturing catches up with latent demand. Therefore, 'today's' decision will influence the type of vehicle that replaces them in 10 to 15 years' time. There could be an existing catenary infrastructure to 'plug into'.

Therefore, costings for all three options need to be in place for comparison with a 30-to-35-year working timescale in order to be practical and achieve the best financial, as well as environmental, result extracted from the time/cost/benefit analysis.

The question as to whether the proposed overhead network of wires in the area of central Lincoln is acceptable is an esoteric argument that should be reflected upon once the cost elements of the three options have been clearly examined. One cannot help feeling that the topography involved with the city of Lincoln location, and arguably, similar locations around the UK, will significantly affect the outcome of the costing analysis comparisons and, therefore, the appropriate decisions made.

There is no doubt that those committing themselves to the investigation as to which option is most appropriate need access to, or better still hold, the specialist qualifications and experience required to make a competent assessment. That assessment may well involve, if not include, the source of funding required to implement the recommendation, whatever it turns out to be.

There are, of course, the socio-economic and health considerations which should be broadly equal in the case of each of the options available, notwithstanding the fact that the short to medium-term will require those considerations to take full note regarding the remaining diesel-based fleet and its morphing into a fully electric fleet for the foreseeable future.

One of the outstanding features mentioned in several of the reports scrutinised is the need for a zealot, as has been mentioned elsewhere. There must be somebody pushing the process forward. Shortening the timescale for decisions to be made is essential whilst ensuring that the detail and data available is cohesive and acceptable. They must build a group of people overseeing the project and ensure the availability of the financial resources required to see the scheme through.

The feasibility study and projected infrastructure reports should go hand-in-hand with the bus operators and their willingness to provide the appropriate vehicles during the initial 10-to-15-year period as well as commit to a subsequent period, perhaps. That long term overview will ensure a continuity of effort that will certainly be pertinent in each and every one of the three options because of the substantial infrastructure costs involved, varied as they may be.

Whichever option is favoured, it will be capital intensive, and that capital cost will need to be amortised over a realistic period of time well beyond the first 10-15 year phase.

Chapter 17.

How Do We Progress the Project?

The basic outline with respect to a value for money long-term solution to the electrification of bus transportation in a city such as Lincoln is outlined below. Other internationally-based reports support the view that an In Motion Charging (IMC) system is the most cost-efficient battery charging system over time when comparing the three options available.

The hills out of Lincoln city make option three the solution of choice more of a statement of preference, taking international experience into account vis-a-vis Vancouver in Canada and Christchurch in New Zealand. Hills are a battery-powered bus's Achilles heel.

This outline focuses on simplicity and engineering ingenuity as well as cost-effectiveness over time.

Simplicity can be achieved by the system having only pressure junctions. The design should ignore the temptation to incorporate selection-instigated junctions to allow variations in the route that the vehicle follows. In-line 'joining' junctions do not require selective operation nor incur the additional cost that is involved.

The passage of the contact heads over the junction automatically provides continuity of contact by dint of the head's pressure on the junction, which springs back once the joining transition has completed. Similarly, crossovers, rare as they may be, could be an essential part of the catenary where a 45° crossover on either side of 90° on an intersection of the catenary can be negotiated without any external input or requirement for disconnection. There are occasions when both are envisaged as being used in the proposed Lincoln design, as illustrated in Diagram 15.2.

On many stretches, and particularly where hills are to be negotiated by the buses, there may only be a requirement for a single catenary track to be on the uphill section. This would save infrastructure costs on the downhill sections due to modern techniques in generating electricity from the vehicle braking system using high-capacity condensers - referred to as 'supercaps' in some technical papers. High-capacity condensers ensure the large part (90% is quoted (1, page 41)) of the energy generated during braking is stored and simultaneously makes certain that a large amount of energy is readily available immediately on the next demand for acceleration power. By ensuring that energy is not lost during braking, there are obvious benefits for the environment in terms of energy recovery in fuel savings.

A further facet of high-capacity condensers is that they prevent sparking when vehicles use junctions or crossover sections in the catenary, which increases the life of the component and reduces any overload that might occur within substations generating the power that the catenary requires to function. Failures in the power supply system are therefore reduced. These components also provide surge protection within the bus electrical system, thus preventing surge power damage to auxiliary equipment when overhead power is interrupted (junctions and crossings) like air-conditioning systems and similarly operated electric motor centred facilities.

The downhill section culminates in a bus stop 'feed' section and a joining junction as described earlier to allow the vehicle to reconnect to the catenary and resume the service.

There may be a need for an operational control centre. The decision-making process in that respect may well rest with the vehicle operators rather than the local authority. One assumes that driver training centres are already in place under the auspices of the bus operators and therefore can be used to upgrade current drivers' knowledge whilst providing initiation training for those new to the job. There will be subtle differences in the operation of the different option vehicles, and therefore, training needs to be taken into account in the costings of the project.

The focus of this publication has been and remains biased towards double-decker format public service vehicles (bus DfT defined). That is not to say that the principles applied do not count just as much for coach bodied vehicles. However, one has to take into account the vehicle footprint per passenger carried and the capacity of the vehicle to do its job. The greater the vehicle length, the more carriageway congestion is created. Also, the more significant an infrastructure change it requires to accommodate the vehicle on the carriageway shared with cars in particular but other means of transportation in general. The vehicle footprint is of paramount importance in consideration of the highway to be used.

The motorcar is not going to disappear from a small city like Lincoln unless and until every home or conurbation can be adequately serviced economically and physically by a bus service that meets everyone's needs. Having stated that opinion, with the introduction of electric cars and the substantial cost implications currently evident in new vehicle acquisition prices, there could be a reduction in demand for cars and thereby, an increase in bus transportation that would be required.

A central control system would need to be available to re-route or divert buses from the central station location, in particular, or any point on the catenary route in general, should an emergency arise.

Specialist engineers and experienced managers will have to be brought in, perhaps from abroad. They will need to examine local features, proposed routes and operational requirements to bring experience and insight to urban planners, architects, and civil engineers regarding the assessment of the project. First-hand guidance will be essential for both the infrastructure specifications as well as, to a lesser degree, the bus operation element of the planned scheme.

Which route or routes would be the initial subject for alteration would also have to be carefully considered in conjunction with the vehicle operator and availability of suitable vehicles. There would have to be a contractual consensus between the vehicle operator to utilise online battery charging and/or in motion battery charging or even overnight battery charging in a central compound because the infrastructure would need to be in place well before the buses were actually put into operation. The option chosen is a mutual choice with long-term implications.

That situation could be advantageous because it gives the operator an opportunity to assess which buses it is going to modify or upgrade and where those buses are going to be drawn from. The options are the rest of the local cohort, another part of its group fleet, or purchased vehicles from another operator; if that is appropriate to the option chosen for the required route operations.

The hybrid bus retrofitting element comes into play not only with regard to vehicle costing but, referring back to the engineering element, with the associated infrastructure implications as well. A cohesive contractual liability to provide service and infrastructure is required to be at the centre of the agreement. Divided responsibility for infrastructure, electricity supply and vehicle operation can be a nightmare as Wellington New Zealand operators found. In short, too little electric current, too small battery capacity and the right vehicles still results in a less than efficient service. The cohesive structure format is outlined in the government paper *Bus Back Better*.(10)

It is probably advisable to have a master plan and a secondary option which accommodate minor alterations or alternative route structures that can also be costed at the same time in conjunction with the master plan. The breadth of the master plan should be the central focus regarding local and district planning documents as they will certainly need to be presented to, in this example, the Lincoln City Council and Lincolnshire County Council, within which all of the main infrastructure work will take place alongside, say, Stagecoach the current bus service operator for Lincoln covering the East Midlands.

What the other papers mentioned in this book will provide is a template for the work outlined above. They will also provide a valuable insight into the providers of specialist knowledge, products and application, which should save a great deal of time. It is to be hoped that the urgency intimated in the *Bus Back Better* paper under the Bus Transport Act and Enhanced Partnership arrangements for local grant funding will speed the planning process to facilitate the proposed transformation.

New innovations develop all the time; however, the costings attaching to most of that development might not be affordable in the short term. A perfectly acceptable compromise arrangement for a 10-to-15-year period with a 30-to-35-year element of pragmatism could provide a stopgap as well as a stepping stone. Any time delay to perfection gives an opportunity for in-depth investigation into progress in the field of electric bus operation (or alternatives such as hydrogen) that will deliver a better outcome for the city. Even the hydrogen option in a hybrid format will benefit from option 3 by way of in service duration.

The suggestion that has been put as a proposed solution for Lincoln City, and other similar environments, utilises technology that is well-known, well-tested and cost-effective. The catenary system suggested takes full advantage of the ability of the proposed buses to connect and disconnect at will. They can be converted to do so in a very cost-effective manner that hopefully provides a speedy solution and implementation of the required technology, compared to waiting for new buses to be manufactured and be available for a particular location.

There is an assertion that the public will not accept the visual impact of the catenary in their town or city required to facilitate In Motion Charging of battery electric buses to operate. A more certain assertion is that bus crews would have an aversion to constantly leaving their bus partway through a journey to disconnect or connect the connecting poles required to provide their bus with electrical current. Whatever the truth of the foregoing perceptions they are certainly based on old technology. The former can be minimised, and the latter can be automated.

Earlier in this publication, we have illustrated the auto-connecting mechanism at the top end of the connecting polls fitted forward of the roof of an In Motion Charging/trolleybus-style vehicle. The in-service activity of the connecting poles, however, is more complex than just connecting the poles to the overhead catenary. The question arises: what happens to those poles when they are not in use?

Historically the bus crew would leave the bus, and, using a pole almost the length of the vehicle that they are operating, they would engage with each connecting pole independently; pulling each pole down to engage with fixed hooks at the rear of the vehicle roof, thus securing each pole until needed to be deployed again. At that point, they would use the same pole, stowed under the bus, engage with the underside of the connecting pole, pull it down and to the side to disengage with the fixed hook and then re-present it to the catenary wire. It was an inconvenience, but a necessary one. However the vehicle stayed connected until its shift was finished because the catenary was a continuous loop.

In later years, the rear of most trolleybuses presented two winches permanently connected by a cord mechanism that would pull down on the connecting poles, disconnecting them from the catenary and drawing them to the fixed hook mechanism for stowage. That was a significant improvement in terms of efficiency and, one suspects, time saved for the bus crew.

In a fully automated situation, two fixed retention hooks present a problem. As anyone who has engaged the fixed hook on perhaps the waistband of a piece of clothing will readily acknowledge the clasp 'hook' has to move in the opposite direction to that desired before the release can be achieved in the decoupling process.

The proposed solution has been designed to automate the process of storing the connecting poles ensuring their release and storage minimises the equipment outlined in the previous paragraphs. Diagram 17.1 illustrates the current format.

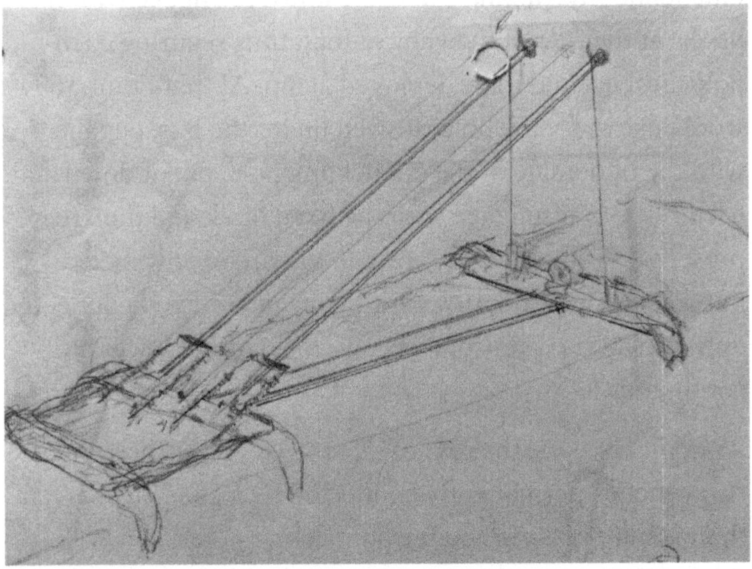

***17.1** Illustrates an automated store and release system.* *[TPO'H]*

On the assumption that the twin connecting poles are connected to the catenary, then both poles can be retracted simultaneously and drawn down by one winch (rather than two, as is common practice historically) into two cam-shaped twin hook mechanisms that, by the pressure of the connecting pole being pulled into the fitting, will close the twin 'hooks' over the round pole sufficient to retain them. That engagement would automatically cease the operation of the retracting winch and lock the equipment in the stowed position.

On deployment, activated by the driver from the cab of the vehicle, the winch would freewheel, allowing the hooks to open and release the connecting poles under the influence of pressure from the forward mounting springs to commence the upward movement towards connection. The winch mechanism would act as a brake to regulate the speed of ascent of the connecting poles in order to prevent damage to the catenary system.

NB A traditional connection retrieval pole could still be stored under the bus to be used in an absolute emergency for disengagement and engagement of the pole retaining mechanism described above.

An image of what was taking place on the roof of the vehicle could easily be supplied through modern technology to the driver's position via a small screen.

The advantage of both connecting poles being operated simultaneously is the combined stability achieved in the structure and the minimisation of the infrastructure required for the stowage/connect cyclic process to be completed. Current flow would be available immediately.

Chapter 18.

At What Cost?

Several mechanisms have been designed and patented that will allow a driver-controlled, pushbutton connection to be employed and facilitate the in motion connection from the bus to automatically connect with the catenary.

The mechanism replaces the current connecting piece at the head of each of the two connecting poles. One example incorporates a winged headpiece that, once connected to the overhead power, swivels away to a 'park' position, well beneath the powerlines. As with any remote operation from the driver's cab there, will be a simple but effective means of securing the poles when they are in the retracted 'out of service' position. That might be achieved using a cam-operated locking hook facility that would hold the two connecting poles independently until the connect button is selected. At this point, the cam hooks would open and the power connecting poles released vertically to reconnect with the catenary to facilitate engagement with the power source.

The simplest way to ensure that both elements are employed effectively and that the retrofitting operation is conducted efficiently and in the least possible out-of-service time for the hybrid vehicle fleet that it will be attached to, is to fit a standard retrofit platform to retaining brackets on the roofline of each model of hybrid bus, specific to the type of vehicle as a complete retrofit unit.

The basic principle of the hybrid bus is that it can easily transfer its motive power from a fossil fuel-generated electrical power source to a battery power source. It should therefore be a relatively straightforward exercise to take power from the retrofit roof-mounted unit. The connecting poles will draw power from the catenary to allow that source of electricity to power the vehicle whilst also replenishing the battery pack of the vehicle it is attached to.

The suggested head (16) and locking devices (17) can be incorporated in the proposed retrofitted platform (18) in an 'off-the-shelf' facility. This would allow any manufacturer or operator of hybrid buses to have their vehicles retrofitted at the most economical cost, coupled with the optimised efficiency of the whole minimised cost/quality concept provided through one standard, mass-produced unit. The only special to-type requirement would be the roof mounting point fittings which would provide the mounting points for the retrofit unit.

The financial argument for the acceptance of retrofitting for the power pole facility is extraordinarily strong. The proposed unit relieves the bus crew of an in-service hybrid bus from exiting the vehicle in order to retrieve the appropriate individual connecting pole placement staff from under the bus to locate the connecting pole to the overhead power supply. This must perforce often take place in inclement weather and applies whether the vehicle is a one- or two-man operation.

The manhandling 'placement pole' is then required to be relocated beneath the vehicle before the journey can recommence. This may be particularly so where a fossil fuel vehicle or an electric bus is involved in transferring from one form of power supply to another. Whatever the weather, the traditional pole connecting/disconnecting process can be considered cumbersome and time-consuming and therefore frustrating to the passengers and crew of the vehicle. That is not to state that if all else fails, it is not still an option to carry forward as an emergency facility.

The advantage of hybrid vehicles in their various guises, as has been discussed, centres on the flexibility of being able to move from one power source mode to another to optimise the environmental and operational flexibility of such a vehicle. The hybrid is also designed to optimise the cost of making that vehicle available to the public. The auto-connect retrofit unit benefits the passengers and the bus crew by minimising the disconnect and reconnection process, thus making the journey a more pleasant experience for both by minimising delays.

The figures that follow are destined to illustrate unequivocally that retrofitting, particularly hybrid buses currently operating on diesel/electric principles using battery power where practical, is a cost-effective and environmentally beneficial adaptation that justifies the capital cost of the catenary infrastructure. The catenary is required to provide an overhead power supply (similar but not the same as that used by electric trains and trams) to be installed along a proposed Public Service Vehicle route. The principle applies equally to a hydrogen engined hybrid vehicle.

Diesel-powered articulated PSVs, commonly known in the UK as 'bendy buses,' have been tried and tested by way of full service over several years in London before being withdrawn, and currently in York and Sheffield, where they are still in use. The problem with operating articulated vehicles is their extended footprint on the road. They add to, rather than detract from, traffic chaos by virtue of their length. Witness a traffic junction with such a vehicle effectively blocking all four entry/exit points due to its road space footprint and restricted manoeuvrability.

There is the added requirement for specialised routes, which in terms of infrastructure cost and detriment to road transport capacity, is intrusive, costly and takes more planning time when compared to the double-decker, single chassis format vehicles that are commonly used in the United Kingdom. There is an acceptance by the public that the current fleet of double-decker units fit UK roads and traffic capacity circumstances more fully than alternatives. The smaller footprint allows all road users to progress more efficiently and faster. The double-decker bus also has intrinsic passenger experience attractiveness, which is often ignored by planners.

It is an acknowledged factor that keeping traffic moving is a critical consideration in road traffic management. Whether the topic is city, urban, or suburban-based road traffic, the public service vehicles must surely fit all three circumstances.

Not everywhere has four-lane carriageways (two in either direction), one of which can be used for the 'Powertrain' style double articulated vehicles operated using the trolleybus principles that one encounters in several European and Asian locations using dedicated 'bus lane' facilities. Most of the countries where they are most used have lower levels of competing traffic due to a relatively impoverished population. Those vehicles would perhaps benefit from the auto-connect unit, but would they economically and ergonomically fit into many cities within the United Kingdom without the need for major changes to the landscape and all that entails?

Historic buildings and narrow roads are part of a British heritage that many accept as needing preservation rather than subjecting to expensive demolition and modernisation, simply to accommodate a new kind of environmentally efficient public mass transport system and the high costs attendant to their introduction and maintainance.

Many eastern European applications are born out of a low-density privately-owned vehicle environment and expansionist urban routing. Funding for Ukrainian, Bulgarian, Romanian, Turkish installations, to name but a few, is often provided by the 'European Bank for Reconstruction and Development Loans' (EBRD) facility as a follow on from the 2013 paper.(1) Installing a new catenary infrastructure will accommodate articulated units. However each of those large articulated units will cost in excess of €1,000,000 each in order to carry 221 passengers (48 seated) on dedicated set route journeys. The EBRD is not an EU organisation; it is owned by 69 countries (including the UK) plus the European Investment Bank.

It will be interesting to reflect on the COVID effect on those mass transit vehicles and the standing passenger configuration.

The foregoing discourse underpins the reason why the principle of an auto-connect winged connector has been expanded in this publication to explore the cost-benefit attributes of retrofitting the established hybrid fleet of double-decker buses in the UK. The principle involved does not preclude its single-decker counterpart application (DfT coach). The concentration on double-decker bus units rather than coach vehicles is purely aimed at the movement of the optimum number of passengers in the most comfortable and accommodating format, coupled with the least negative impact on other road users.

The analysis starts with the cost of new double-decker hybrid buses. That factor exemplifies the cost of 'renewing' the current fleet of 10-15-year-old vehicles as an alternative strategy.

The following tables utilise historical data to note the decrease in value over time of a vehicle in service and the expected lifespan of, say, 20 to 25 years from its introduction into service. The vehicles considered in the tables travel in excess of 1,000,000 miles in their lifetime. Much happens during those million miles in terms of cost, particularly with the battery element central to the hybrid and battery only bus concepts, as previous chapters have explained.

The battery life expectancy of a hybrid vehicle may be as much as three to four years in regular service. (1, page 67) However, at that point, a significant capital cost is encountered to replace the battery units, which by then have no real recycling value of any note. The cost of transporting redundant batteries for disposal reportedly exceeds the scrap value of the unit. Currently, it requires the 'scrap' to be transported from the UK to Belgium with the environmental penalty that the HGV carrying the waste generates in carrying out the journey. It is a falsehood in extremis.

The overhead power supply addresses that problem by extending the life of each battery thanks to the trickle charging facility provided whilst the vehicle is in motion and performing its required service (i.e. moving towards its passengers' destinations).

It is said that battery life can be extended by a further two or three years if the overhead catenary supply is brought to bear to give slower-measured charging whilst the vehicle is still in active passenger service and on the move. The battery is then used solely for city centre or specialist route capability where the overhead catenary would be inappropriate (perhaps interfering with Christmas or other holiday decorations, or merely inappropriate because of other factors).

The cost-effective nature of the retrofitting of the vehicle takes on an additional dimension which makes it even more capital effective. The figures in the table show the enhancement to vehicle value and longevity of service compared to the purchase of new vehicles with that same overhead power supply capability together with the alternative options and therefore performing an identical service.

Retrofitting will create between seven and eight trolleybus style In Motion Chargeable PSV units in return for the same capital cost as one new hybrid vehicle. [The author believes that the production of Euro VI hybrid vehicles must be maintained and, if possible, escalated to meet the environmental targets that have been set, particularly in respect of the sub-Euro VI fleet comprising 15,600 bus units alone (i.e. ignoring the coach vehicles)](6).

The capital saving will more than justify the cost of the catenary infrastructure to be installed in order for the vehicles to be brought into service. Contemporary research, in Europe in particular, but across the world in general, indicates that the public who travel on electrically powered vehicles love the quiet, smooth operation of trolleybuses compared to the noisy, vibration accompaniment that emits from diesel-powered transport. Those who live adjacent to the routes on which the electric vehicles operate also experience better health. City-operated hybrid PSVs currently revert to diesel power perforce because of the short duration battery life in operation. (1) The catenary resolves that deficit of electric power.

The European Union report of 2013 (1) is as pertinent today as it was eight years ago because there has been so little progress in extending battery life to a practical and meaningful extension of duration.

That report argued then, as this publication argues now, that the trolleybus with its history going back to the 1900s has proven itself worthy of not only consideration but of implementation. The double-decker trolleybus should therefore be re-established in the United Kingdom as a principal form of public transport. The In Motion Charging bus can operate to full 0% environmental hazard in urban and city environments whilst being able to revert to ergonomically acceptable fossil fuel application in suburban and rural areas where any environmental detriment will be minimised.

The ability to connect, disconnect and reconnect to overhead power supplies 'at will' anywhere on the route where there is a straight road of at least 25 feet in duration is a huge plus for the trolleybus hybrid adaptation and its reinstatement on our PSV routes. The figures speak for themselves.

The assessment of costs in respect of a variety of bus types compared by propellant and time in service is an interesting undertaking.

The chosen subject vehicles include the Dennis Enviro 400, which is a hybrid bus that was introduced in 2005. That vehicle is compared to the Volvo/Wrightbus Eclipse Gemini II from 2009 and the Transport for London New Routemaster hybrid bus introduced to service in 2012. Each has somewhere around 1,000 plus buses operating in the UK, and therefore there is a second-hand market value attributed to them. (3) They also have a history of use available for assessment purposes rather than theoretical assumptions, so often proven to be incorrect or at least fall short of their assumed performance expectations.

The tables are intended to present the values of a battery electric bus of no particular manufacture and generally priced at 2015 values, together with the hydrogen bus produced by BYD UK of Chinese origin at £500,000 each, reportedly purchased by Transport for London (TfL) to bring the discursive analysis up to date in September 2021.

The tables consider the cost of the vehicles as new and establish the power output from each of the vehicles. The cost of running the vehicles does vary because, of course, the trolleybus requires an infrastructure by way of the catenary, which is not attributable to any of the other forms of motivation, per se.

However, those other vehicles considered in the comparison do have their own speciality infrastructure requirements, such as fuel pumps, roadside charging facilities, hydrogen storage, and so on. Therefore there are infrastructure costs attributable to all those vehicles that will affect the comparison data. It is not within the gift of this publication to bring those costs to the debate in detail; however, the referenced reports do attempt and, on occasion, specify cost comparisons that arrive at verifiable and therefore acceptable conclusions. (1, page 67, table 4.1)

The authoritative tables produced in the 2013 discussion paper by the EU entitled *Take-up guide for the Replacement of urban Diesel buses by Trolleybuses* (1) do assume to portray those figures albeit for a variety of configurations of subject vehicles, including the power train with a higher initial capital cost from which it still comes out ahead of its fossil fuel counterpart.

The following table shows the life of any stock vehicle in years, given due consideration for miles accomplished, chassis life and wear appropriate to the battery renewal period, based upon standard driving in moderately undulating conditions and how that battery life can be affected. A hilly environment will apply some notable disadvantages where pure battery operation is concerned. There is a lack of power sufficient to negotiate undulating terrain at a reasonable rate whilst benefiting from the advantages of retardation on the downward slope in delivering recovered power to the battery system is often quoted as a balancing consideration. It is not a balancing consideration in any context.

The primary consideration that is believed to be unique to this particular analysis is comparing cost when considering the vehicle depreciation (progressive devaluation of the asset), coupled with the added value that is attendant to the retrofitting. Allowing overhead access to electrical power and a concurrent internal refit to the hybrid vehicle fleets bolsters their value over time and extend the vehicles life and facility by some considerable margin.

The advantage of retrofitting is concerned with the almost immediate operational advantage in environmental and capital cost terms as well as the reduced vehicle running costs over the remainder of its useful life in service.

How long it will take to create and perfect the design of hydrogen power in public service vehicles (and heavy goods vehicles as is the application now in the USA) remains to be seen. Where will those vehicles be deployed? Aberdeen may well provide the answer on reflection in a few years when its newly acquired fleet of hydrogen propelled buses has been evaluated in service.

Retrofitting existing hybrid buses will go a long way to eliminating the detrimental effects of diesel engine operations in cities and heavily populated urban areas, particularly in the 'north of Watford' environment where older bus fleets prevail. Retrofitting will give a 'time' advantage to improving environmental conditions and reducing the detrimental effects of sub-Euro VI diesel engine operations in cities and heavily populated urban areas.

The upgrading of in-service vehicles will give a time advantage to improving the environmental conditions whilst mass production, at reasonable capital cost, is created for the alternative. This may well be, with the expected life of the catenary infrastructure at 30 to 35 years, a dedicated trolleybus-style hybrid or fully-electric In Motion Charging bus entering service as an interim, if not permanently acceptable, long-term solution on some routes.

The figures assume an industry standard of 50,000 miles per vehicle per year in service with a twenty-year life creating a 1,000,000-mile chassis lifetime which could well extend for five or ten years beyond that statistic; in itself, making for a remarkably interesting comparison table.

The discounted value of each bus pre-retrofit in the first five years amounts to between 10% and 18% of the original capital cost, with a leaning towards the latter.

Assessment of costs in respect of a variety of bus types by propellent and time in service

		A	B	C	D	E	F
Table 1					battery		
		hybrid 2005	hybrid 2008	hybrid 2012	electric 2016	trolleybus	hydrogen
make / type		Enviro 400	Eclipse GEMINI 2	New Routemaster	Optaire elect		
Country of manufacture		UK	UK	UK	UK		China/uk
name of manufacturer		Alex/Dennis	Volvo /Wrightbus	Wright bus			BYD uk
The cost of the vehicles		£132,000	£220,000	£354,000	£380,000		£500,000
power output		138kW/185hp	194Kw/260hp	130kW/174hp	Optaire		
The cost of running the vehicles					£80,000/4 yr/batt		
infrastructure required		diesel fuel	diesel fuel	diesel fuel	charging point		
				charging point			
The cost of the infrastructure to run the vehicles		fuel pumps	fuel pumps	fuel pumps	electric charging	cantenary	fuel installation
				electric charging			hydrogen source
The life of a stock vehicle - years/miles		25yrs /1.5 million	25yrs / 1.5 million	25yrs / 1.5 million	?	30 yrs / 2 million	?
battery renewal -years /miles				3yrs/200,000 miles	4yrs/200,000 miles	n/a	
The latent value in the existing stock of vehicles	5yr	£120,000	£180,000	£255,000	£202,000	n/a	
	10yr	£73,000	£110,000	£175,000		n/a	
	15yr	£30,000	£45,000				
The cost of retrofitting the existing stock		£50,000	£50,000	£50,000	£50,000	dedicated	n/a ?
The extended life of a retrofitted vehicle		10yrs	15 yrs /	25 yrs /			
age of vehicle at retrofit				20 yrs /			
Twenty year life to attain 1 million miles.							
		Gemini	Enviro	hybrid 2005	electric		
percentage of origin	5yr	81.82%	90.91%	74.00%	74%-£80000	94.00%	
	10yr	50.00%	55.30%	47.00%	?	73.00%	
	15yr	20.45%	22.73%	?	?	38.00%	

***18.1** Table 1 assessment of costs.*

At year ten (which accounts for a significant number of fleet buses in service in the year 2020), the reduction in value is between 45% and 50% of the original cost.(3) These are pertinent figures where retrofitting is considered because the cost of retrofitting and creating the catenary required to run the vehicles can be appreciably less than awaiting the delivery of newly manufactured vehicles (with little extra service benefit in terms of the lifetime use of the replacement vehicle).

Why buy a replacement vehicle? The overall time/cost/benefit washes out for the retrofit vehicle between year five and year ten.

Even at that extended point, year ten, the infrastructure, the catenary, and the associated power requirements still have a further 25 years of service available to power replacement newly manufactured vehicles at that time.

In terms of value for money, that proposition is surely unbeatable.

The cost of retrofitting with the retrofit platform will be between £17,000 and £21,000 per vehicle.

If the 10- or 15-year-old buses were refitted internally, giving them a further 10 or 15 years' service, that cost would be between £15,000 and £18,000 in 2021 terms per vehicle.

This combined retrofit and refurbishment can be achieved at an overall combined cost of circa £42,000, including the electrical adaptations that may be required to accommodate the retrofit platform. Taking the amortisation process that bit further, we can assume an amortisation period of ten years for the retrofit unit (although 20 years' service is the realistic expectation). The remainder of the original amortised period for the highbred being retrofitted with a local authority dividend assumption of yearly increments gives an overall, realistic base for comparison appraisal.

Table 2 illustrates the cost-effectiveness of the vehicles after retrofitting, even on these short term assumptions.

Table 2	AMORTISED COST OF SUBJECT VEHICLE AT DATE OF PURCHASE					
	Diesel 2005	Hybrid 2009	Hybrid 2012	Electric	Trolleybus	Hydrogen
	Enviro 400	Eclipse GEMINI 2	New Routemaster	Optaire elect	?	HYD 001
	UK	UK	UK	UK	?	Aberdeen
INTEREST	Dennis	Volvo/Wright	Wright bus	Optaire elect	?	Wright bus
RATE	£132,000	£220,000	£354,000	£380,000	£390,000	£500,000
1.00%	£3,649	£6,082	£9,787	£10,505	£10,782	£13,823
2.50%	£4,214	£7,023	£11,300	£12,130	£12,449	£15,961
5.00%	£5,258	£8,764	£14,102	£15,138	£15,536	£19,918
	Diesel 2005	Hybrid 2009	Hybrid 2012	Electric		
RETROFIT	Enviro 400		New Routemaster	Optaire elect		
COST	UK		UK	UK		
	Dennis	Volvo/Wright	Wright bus	Optaire elect		
	£42,000	£42,000	£42,000	£42,000		
1.00%	£2,327	£2,327	£2,327	£2,327		
2.50%	£2,694	£2,694	£2,694	£2,694		
5.00%	£3,370	£3,370	£3,370	£3,370		
Sum@2.5%	£96,016	£208,380	£370,285	£539,086	£497,974	£638,429

***18.2** Table 2. Amortised cost analysis per vehicle type over a 20-year period.*

As has been established, the cost of a simple catenary installation could be anything from £1 million - £2 million per mile of two way, outward and inward movement facility. Each catenary will be capable of powering, say 4 to 6 vehicles per service mile, given a twin wire outward and return overhead power facility. Once the catenary is in place, it can provide cost-efficient, low maintenance service for up to 35 years.

***18.3** A catenary view at The Trolleybus Museum, Sandtoft, North Lincolnshire. The proposal simplifies this.*

Given the economic conditions prevailing in 2021, coupled with the urgency of the environmental necessity to reduce not only CO_2 but nitrogen particle effluence contaminating the atmosphere, the retrofitting of current hybrid vehicle bus fleets and the installation of an overhead power provision capability has no credible alternative in capital including amortised utility, nor time-to-install terms.

One of the advantages of retrofitting a hybrid bus is that when it disconnects from the overhead cabling to proceed through a suburban or rural area in order to complete a scheduled journey, the configuration of the bus for energy efficiency is such that it reuses the recuperated energy from braking for traction purposes which in itself creates a 20% saving by way of fuel economy. Therefore, the fossil fuel engine is not all bad in that suburban and rural area environment. (1)

That same facility is available to save electrical consumption when the vehicle is connected due to the higher energy efficiency of the recuperated energy braking mechanism. That same mechanism provides a saving in respect of the braking mechanism employed on the heavy goods vehicle fleet, thus reducing the amount of brake dust emitted into the atmosphere from normal fibre-based brake pads.

In this analysis, it has been assumed that a public service vehicle unit operates for somewhere between 310 and 330 days per year. On that assumption, the life of a battery in any battery-driven vehicle would be circa three years. This is largely due to the number of charging cycles involved in the vehicle's operating environment. (1, table 4.1).

A further financial implication relating to battery-driven vehicles is created because of their operational duration restrictions, which results in a larger fleet of vehicles being required to achieve a given level of service.(10) The service requirements of the fleet as against diesel or trolleybus applications which are available between 'refuelling' is probably another 50% in the case of diesel versus all battery power and unlimited from the overhead facility with reference to a conventional trolleybus which is 'fuelled up' in service. The result is, of course, that the fleet operator requires 20% more vehicles with the attendant capital cost involved in that equation. The alternatives to In Motion Charging buses are expensive vehicles. (10)

As a summary of battery-driven vehicles:

a) There is a significant uplift in cost as more vehicles are required to carry out fleet functions.

b) Battery life is short because of high-density charging. A further capital cost, occurring every three years.

c) When they have passed their useful limits, the batteries are not economically recyclable to any ecological, financial or environmental advantage.

In the longer term, the material availability required in the manufacture of the number of batteries potentially required in the event of a complete international conversion to battery power alone may be a problem. The basic materials employed in the batteries are a finite resource, and that resource is mainly sourced through China although mined elsewhere.

Chapter 19.

A Source of Expertise

It is said that seeing is believing. In many ways, that is the truth of many people's means of understanding. With that in mind, the following question is posed: where can we see double-decker buses operating from a catenary?

The answer to that is the Trolleybus Museum in North Lincolnshire which operates a timetable of trolleybuses on a regular basis during its open days. Sandtoft Trolleybus Museum is a well-established facility that operates trolleybuses on a fully established catenary from which much useful information can be drawn and used effectively.

Similar facilities are located at the East Anglia Transport Museum Carlton Coleville, The London Transport Museum and Crich Museum in Derbyshire among others. Wherever you live there should be a facility not too far away.

If seeing is believing, then hearing is understanding. That is achieved by talking to experts, and by that, one means experienced engineers dealing with exactly the style of vehicle that employs the fundamental skills of In Motion electrical power and has done so over many years. Indeed, the Sandtoft Museum proudly hosts the trolleybus that evidences the 1986 attempt by South Yorkshire Transport to reintroduce the trolleybus in Doncaster, South Yorkshire, which is why it features here.

This vehicle provided an extremely exciting possibility for the revival of not only Doncaster but Rotherham and Mexborough trolleybus services (In Motion Charging vehicles) in much the way that has been proposed in this publication for the city of Lincoln. We do not need to stop at Doncaster. Huddersfield, Bradford, Newcastle and many other northern cities are prime examples of where the relevant science has provided benefit in the past and can do so again in the future. It can work.

The Sandtoft Museum is a working centre for 60 to 70 passenger capacity vehicles. It epitomises the efficiency and longevity of the proposed reintroduction of the catenary as a power source and the double-decker trolleybus with its new mode of operation as a hybrid vehicle combining battery power and mains operation into service in the UK.

Access to experienced engineers, administrators, and civil contractors with knowledge of all aspects of infrastructure and trolleybus operation could not be more centrally placed. Whilst some of the most up-to-date internal technologies will need to be sought elsewhere, there is no better place to study the basics. Due to the expansion of the museum, access to the current cost of infrastructure installation practicalities is available from those that serve the museum so well.

This publication owes much to Dave Chick, the editor of Trolleybus Magazine, (3) which constantly updates its readers on the transformation of the trolleybus systems worldwide on the ebb and flow of pertinent debate concerning the operation of vehicles predominantly produced in Europe, Russia and China. Up to date as well as historical information is readily available from its pages, giving an insight into aspects of operational interest that is not readily available elsewhere. David is also heavily involved with the Sandtoft Museum.

The catenary system at Sandtoft operates in exactly the same way as any other catenary anywhere in the world operates. Therefore, it provides first-hand information for comparison and research purposes available to any proposed installation and the local authority and/or government representatives in search of factual comment, both historical and current.

If Sandtoft and a similar museum at Crich hold an illustrative key to In Motion Charging, then Transport for London can provide equal commentary on a relatively local level (i.e. within the United Kingdom) with regard to the operation of centrally charged Battery Electric Buses (BEBs).

That same organisation provides up-to-date and meaningful statistical analysis for diesel hybrid buses operating with Euro VI diesel engine vehicles combined with battery power by way of the new Routemaster. (10) Going further afield to Aberdeen, there is an opportunity to gain first-hand knowledge of the operation of hydrogen-powered vehicles and their foibles.

By using UK-based comparisons, one can extrapolate the results of the other technical and research papers included in this publication. The main difference in all the comparators is the financial currency. However, for the most part, all of the research that has been examined in relation to papers delivered post-2013 and up to 2021 concluded that despite the perceived, though unproven, dislike of the overhead wires, the catenary system displays the time/cost/benefit advantage over the alternative, battery operated random charge and central charge options by a significant margin.

A visit to Sandtoft Trolleybus Museum will soon quell the fears over the unsightly intervention of a catenary to make 'In Motion Charging' a viable opportunity to charge onboard bus batteries slowly, thus extending their life span and keeping more vehicles in service for a longer period of time than any alternative available today. The band of experts, as experienced personally at the Sandtoft site, will readily confirm that affirmation.

Chapter 20.

The Way Forward

This publication is by way of an insight into the possibilities of reintroducing trolleybuses into public service in Great Britain. The argument is compelling provided that there is an objective understanding regarding the assessment of the amortised application of the proposed capital cost factors.

The primary task is to find an enthusiastic local group that will take the ideas presented here to those in authority, utilising the arguments presented here. This information, coupled with the unassailable belief and understanding that 30 years of experience in the UK with the trolleybus through to the 1960s, presents proof positive that the expenditure is worth the benefits achieved in the UK. Many international trolley bus applications have been in service for over 70 years and are expanding their operations, Vancouver, Canada, and Italian cities among them.

Significant individual sources of expertise can give primary advice regarding the catenary installation, which is essential in making the in-motion power transfer concept using overhead power lines a reality. That expertise regarding modern installations exists in the UK, across Europe and Asia.

Are the new facilities that far removed from the installations of yesteryear that were dismantled in the United Kingdom in the mid-1960s?

The technology has advanced significantly; however, the basic infrastructure remains the same in principle. Therefore, the effect on the health of the public in general and the reduced detriment to buildings remains the same. The catenary powered In Motion Charged battery bus and modern trolleybus operation is an inherently clean, noise-free technology that has moved on to improve an already efficient mode of transport.

19.1 *An image of the past - a UK trolleybus of conventional style.*

Europe has used trolleybuses and expanded its fleet of vehicles historically in a single-deck 'coach' format. They have extended that format with articulated vehicles to form 'power trains,' mass-transit units on which much of their system growth depends.

***19.2** The three-car articulated 'powertrain' – note the relative size of the van. [Hese]*

The unit is 20 yards long. (60 feet/18 metres) A cricket pitch wicket to wicket is 22 yards. (20.12 metres)

Evidence suggests that the powertrain format is not generally appropriate for application in the target cities of Great Britain along with urban areas. This is because it entails major expenditure on demolition, road widening and dedicated traffic lanes. It also increases congestion and construction pollution, which is in doubt in achieving a practical, environmentally driven solution. (19)

Perhaps Britain has more cars on the road and heavy goods vehicles per capita than the continent. Great Britain certainly has less space in its urban configuration, with notable exceptions. There are significant financial and environmental implications and savings to be considered, as well as practical historic preservation.

The conversion of double-decker buses already in service, hybrid buses in particular, involves minimal alteration to the mode of transport currently available. That has to be a serious consideration. These buses are already partly powered by electricity; therefore, introducing that electricity from an overhead catenary to charge the batteries embodied within the vehicle makes sense in so many ways, battery life extension for one.

Financially, the argument adds up because at least seven operational In Motion Charging vehicle units can be created through a retrofitted facility being applied against the cost implications for every new bus that could be purchased. Even that claim is a conservative view of the savings that can be achieved regarding the capital outlay on the vehicles involved.

New vehicles are required to replace the thousands of below Euro VI standard diesel buses currently in service. Manufacturing output has to increase dramatically to achieve that. The retrofitting and refurbishment are achieved separately from the manufacturers by specialists in the field that are used to doing that work, having done it for many years. They will need to expand.

***19.3** An artistic impression of a Transport for London New Routemaster with 'In Motion Charging.'[Innovate]*

Environmentally, the overhead cabling, once installed, has a useful life of between 30 and 35 years. Maintenance is minimal. Therefore, it is cost-effective to bring the refitted hybrid buses into service. This extends existing vehicle lives by 10 to 15 years, whilst new dedicated hybrids are manufactured and produced with an expected lifespan of 25 to 30 years themselves on the then-existing catenary available. These may be Diesel Hybrid, Battery Hybrid, or Hydrogen Hybrids.

The creation of an environmentally friendly inner city and urban transport system using the existing road network has its attractions. A new catenary infrastructure connecting important destination points should also attract people to return to the city or urban environment to live rather than pushing residents out into the suburban or rural areas, as the current contaminated air possibly does. The catenary routes proposed are envisaged initially to facilitate high usage routes connecting large residential estates with the main work areas, secondary schools, hospital facilities, the rail network and central bus hub as stated earlier.

If that cleaner air is accompanied by efficient, regularly running, quiet, vibration-free, pleasant to travel in vehicles (which the trolleybus by public accord is acknowledged to be), then that in itself is a good reason why the proposed system should be reintroduced to Great Britain and Northern Ireland's transport system without delay.

Asthma, emphysema, bronchitis and other breathing impediments are prevalent within inner-city areas. Whilst battery buses appear to be ideal in resolving the air quality causation, they are expensive to purchase, and their route duration is restricted. Also, journeys are interrupted, of necessity, to 'refuel' by means of 'fast charging units' situated along the route, creating frustration with passengers and crew alike, quite apart from the bulky nature of the recharging points obstructing the footpaths and the attendant capital cost of their installation. (1, table 4.1)

Assessments from the EU 2013 *Replace Diesel with trolleybuses* paper (1) indicated that while the speed of travel was not of overall importance to the travelling public, the journey's progression certainly is, along with comfort, safety and ecological benefit. Interestingly, the fact that older units might be employed was not a major impediment to the public's use of that form of transport which reinforces the retrofit program. In that respect, the retrofitted hybrid trolleybus wins hands down. (1)

Whilst there is a move to get private cars and other vehicles off the roads, particularly in the city and urban environments, one does not need to make an enemy of the motorist by placing 20-yard (60-foot/18 metre) long vehicles, prone to regular stopping in the carriageway that the other vehicle users are attempting to use. Surely the design objective must be to get the vehicles that the passengers board to travel the furthest possible distance to take that passenger, unimpeded and comfortably, to a destination that may be quite far away from their starting point in a suburban or even urban environment. The retrofitted double-decker hybrid vehicle is more than a match to achieve that desired outcome, particularly if the proposed through city dual destination mode of operation is adopted.

A 60-foot-long (18 metre) double articulated vehicle on a two-lane highway is hardly endearing to the public at large. The vehicle creates congestion unless a special provision is made for it with the attendant (and seemingly unnecessary) infrastructure alterations and inflated cost of the exercise. Where it does fit, perhaps, is in the Cambridge style operation. Yet even there, the bus is a worthy competitor.

Flexibility is impaired purely by the length of the unit. There is, perhaps, a justification when comparing the mode of operation to a tram. Large cities have a tram philosophy in place, and there is no reason why one mode of transport, where the infrastructure cost and passenger movement is justified, should not work together as complementary units as the railway and current bus system is capable of doing.

All current types of vehicles have to live with each other. By contrast to the 'powertrain' principle, a 25- to 30-foot-long double-decker vehicle is a lot easier to overtake and therefore 'live with' on a normal highway than a powertrain. It is of itself more manoeuvrable to skirt roadworks or breakdowns, parked vehicles etc. Such a vehicle also has the duration when using overhead power lines for a part, if not all, of the journey to be undertaken to accomplish both the distance and safety of travel for everyone, passenger and general road user alike.

Another aspect of the articulated 'powertrain' vehicle concerns the collection of fares. The incidence of non-payment on such systems is endemic with the multi-door/multi-carriage configuration. Passengers should contribute appropriately to the public service transport system. The bus format achieves that in its current manifestation. The travelling public are used to the format. Contactless fare collection is becoming commonplace, as are advance purchase tickets from machines.

The advantage of the catenary and the retrofit unit with particular focus on the connecting head and the auto engagement facility that it incorporates is that the catenary power supply does not have to be continuously available as it was in the old trolleybus days. The hybrid vehicles, by design, will have a diesel/hydrogen fuelled engine and/or a battery pack alternative power source to use when disconnected.

The battery pack serves a dual purpose in that it provides power when the conventional engine is not required whilst it also replaces the ballast that is essential to a double-decker configuration to ensure the inherent stability of the vehicle itself. The battery pack assists in lowering the double-deck bus centre of gravity. Therefore, the battery is not a 'wasted weight,' it can be a positive feature serving that valuable dual purpose. It also gathers and stores power from braking, thus saving time and cost by recycling energy.

The In Motion Charge bus that is envisaged might start from a city centre on battery power alone and then connect to the catenary at an appropriate point where the overhead lines do not impede street decoration or historic building considerations. The “In Motion Charging” concept is proven technology that is being used in many countries.

That ‘connected’ journey can continue with the battery being charged from the overhead lines at a leisurely pace whilst the vehicle is still in motion. There is no passenger impediment or journey time loss. What is the future trolley profile?

19.4 The production trolleybus of tomorrow may well look similar to this.

In winter, the electrically-powered vehicle heating system can be maintained for the ‘connected’ element of the journey, including at bus stops where a diesel engine might otherwise perforce be ‘cut out’ to save emissions and fuel consumption. The battery takes up that power requirement in the absence of any external supply. A similar power conundrum applies in summer with air-conditioning and the power required for the facility. The catenary provides that extra power required for passenger comfort.

One does not have to be frugal in either circumstance when the overhead power lines provide sufficient power to drive the vehicle together with the ancillary electrical requirements. For a battery-powered bus, that is an impediment and a constraining feature. Even for a hybrid bus, as the new Routemaster in London has demonstrated, it impedes the vehicle from running on battery power alone for any extended duration. Battery breakdowns are numerous, as shown in the Wellington, New Zealand and Brest, France experiences.

When the hybrid vehicle has reached a lower density of population coupled with lower density traffic movement, it can easily change to diesel power (disconnected from the catenary) and travel through villages and rural areas to complete a journey to distant route destinations without the consolidated detrimental effect attributable to a city centre environment and multiple vehicle emissions. The subject vehicles own emissions are reduced because of the speed of the vehicle between stops and the duration of each transient stage between stops. That compares favourably with the retarded progress in urbania. The engine is required to do more work due to frequent passenger stops, traffic signals and low-speed progress, which therefore produces more toxins.

In fact, the retrofitted hybrid vehicle could run from a rural starting point, through a city centre and out to the other side of the city to a further rural area without the impediment of any necessary stops to recharge batteries thanks to the ability to reconnect to the overhead catenary through the built-up area. The vehicle becomes of greater utility and therefore gains enhanced earning power by remaining in service.

The connecting device and separate automatic parking element for the overhead poles ensure that everything can be controlled from the driver's cab with minimum inconvenience to the travelling public. Surely that is a worthwhile and achievable goal in its own right?

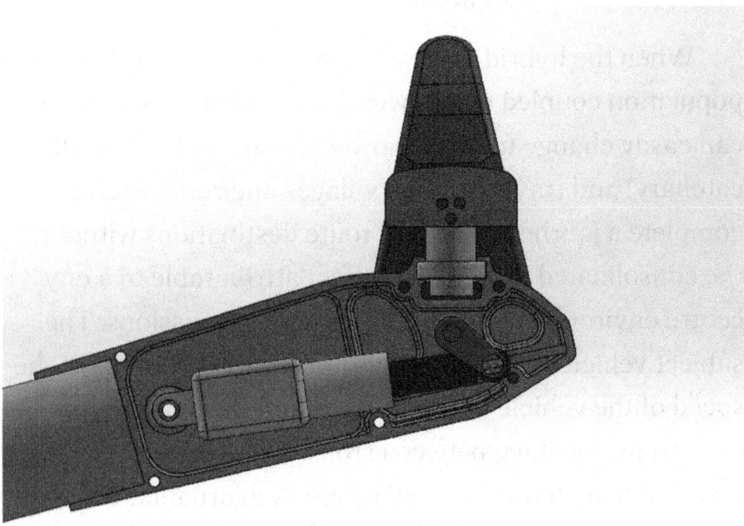

***19.5** Multibus™ connecting head unit deployed to connect.*(16)

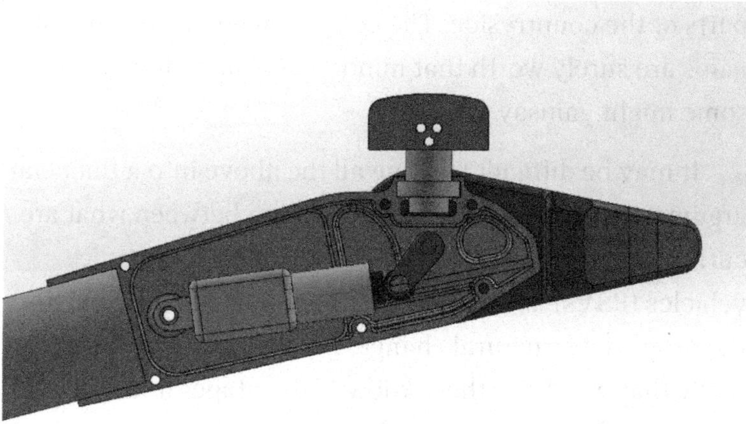

***19.6** Multibus™ connecting head unit in 'service' mode.* (16)

Catenaries can be unobtrusive, multi-purpose and almost invisible. The structure required for suspending the overhead power wires can be a dedicated unit - a slim streetlamp-style structure incorporating an outgoing support beam that will suspend one and possibly both inward and outward travelling power wires.

Many cities are renewing their street lighting systems to more efficient low power facilities through which a dual purpose can be achieved with the supporting uprights providing a lamp standard as well as the catenary support arm. A variety of public service facilities can be supported by way of public address systems, radio signal distribution, broadband projector/receiver units and so on. The overhead wires adopt normality in the same way that overhead telephone cables and power cables do in many parts of the countryside. The ecological and environmental gains are surely worth that minor visual impediment that some might gainsay.

It may be difficult to bring all the above into a financial argument to enable a direct comparison between what are currently predominantly a diesel fleet of public service vehicles (PSVs) and the alternatives, both new and old, to the proposed structural change. But are all the alternatives really that new? Are there known advantages and vices that can influence the normally protracted investigative and analytical processes?

As this publication has endeavoured to clarify, the hybrid principle dates back over 100 years with the introduction of electrically powered vehicles equipped with an onboard generator. Trams have shown their benefits and impediments over as many years and certainly have a place in mass transit, but at what cost in terms of financial, ergonomic, and economic disadvantages? These impediments are historically available for analysis and have not altered a great deal.

Flexibility and cost considerations are two parameters that have relevance in public transport provision, and a third is user cost contribution. There is a capital cost that requires amortisation over a time period. That must surely take account of the operational cost of any vehicle and the work it is expected to do within that infrastructure to provide a public service over time.

Historically, and currently, bearing in mind that circa 44,000 trolleybuses operate worldwide, some of them exceeding 30 years vintage, the trolleybus has proven that the capital cost of the infrastructure that provides overhead power to the vehicles concerned is more than value for money when amortising the inherent capital cost over 25 years' service(1)(2)(10). Even adding in the operational and maintenance costs over that same period proves cost-effective against its pure diesel operated counterpart.

It is also probably worth remembering, although wholly inappropriate to this discussion, the old lead-acid batteries in 1960's trolleybuses were 100% recyclable. That cannot be said of their modern lithium or other 'modern' counterparts.

The cost of a catenary infrastructure can vary not only from country to country but from location to location. The figures gleaned from recent copies of Trolleybus Magazine, (4) which keeps a 'weather eye' on the international application of trolleybus operations, including catenary installations, indicates that costs can vary from one or two million pounds per mile, including substations for a dual installation (incoming and outgoing) and twenty million pounds per mile for the more elaborate powertrain style service introduction to service.

With those figures in mind, it is essential that a committee tasked with assessing the introduction of electric vehicles and interim formats engage specialist architects and engineers to carry out the feasibility study for their local authority or government departments to ensure that those individuals understand and have experience of the infrastructure that they are endeavouring to analyse and implement. They need to have experience of catenary technology.

It is to be hoped that this publication and the promotion of the conversion of the existing fleet of hybrid vehicles to an In-Motion Charging configuration by retrofitting hybrid fleet vehicles through the fitting of a dedicated standard format unit incorporating an auto-connect and retraction stowage system will keep the cost of the vehicles involved to a minimum. By this means, it will allow that saving in precious capital to be invested in the catenary infrastructure necessary for those vehicles to run effectively and cleanly within a modern city as a focal point in the UK of tomorrow.

Perhaps the most important but least talked about consideration is how much electricity is required for any and each application. Also, where will that electricity come from? Hydrogen-powered buses require a substantially environmentally adverse process to produce their fuel. If the whole bus fleet was converted to battery operation based upon overnight or block charging, the national grid would have to expand to produce three times its current output. (27)

A sensible compromise in the form of the In Motion Charging-based hybrid diesel VI based bus provides the best of both worlds. A lower electrical demand coupled with the flexibility required for universal bus operational requirements. The environmental improvement is significant and can be established in the shortest time scale with significant cost advantages.

A further consideration is nuclear fusion-generated electricity, using small local generating plants similar to those that have been in production since the 1950s to power submarines and other vessels. WE must not confuse nuclear fusion with nuclear fission. Rolls Royce must surely be able to adapt the marine 'engine' to onshore applications. The smaller units can be mass-produced and in service within a shorter time scale than the traditional power station format. A secondary benefit could be a replication of the energy-saving application along the lines of that employed by Battersea Power Station all those years ago providing heating to local homes and facilities. It is just a thought. Luddites beware!

Chapter 21.

Ownership and Funding

Many commentators on the public service vehicle provision funding debate compare the deregulated rail system franchise architecture with the possible funding format for road transport. Nationalisation is always an agenda item. The inherent problem with nationalisation can be the lack of competition and, therefore, the lack of progressive achievement and upgrading of the subject matter. "Need is the mother of invention."

The national bus company that took over British Leyland and the Bristol bus company, along with all the attendant coachbuilders such as Eastern Bodywork, may provide a classic example. Vehicles made under license by independent companies such as Volvo and DAF incorporated modifications to their licence-built engines and body structures, increasing fuel efficiency, speed and comfort, eventually putting their host licensor out of business.

It appears to be the story of British industry. The UK led the world with train technology, and now we import them or build them for foreign owners. The British motor industry manufactures more cars and makes of car than any other comparable nation. Most of the manufacturing units are owned from abroad. Arguably nationalisation failed to help the process of UK motor manufacturer ownership, yet we have the diversity of available products.

Naturally, there are funding reasons. Of course, there are attempts to save an industry from itself. It is not helpful to deny that. The question is, how does one maintain competition that encourages development and progress in the marketplace whilst providing value for money for industry and its customers? It is surely a fine balance.

The rail network franchising system failed for a variety of reasons, most of which are highlighted in an interesting paper by Nigel Hawkins, an infrastructure and renewables specialist at Hardman and Co, specialising in the energy sector. He is also a senior fellow of the Adam Smith Institute. Hawkins's paper, *The Failure of Rail Franchise,* (19) is a modest overview of a system claiming to be privatised in the mid-1990s, which has now come to a major rebranding following the Department for Transport decision to cosmetically amend the franchising system that went hand-in-hand with that privatisation initiative. Was it ever a fully privatised system? This question appears open to conjecture. (25)

The franchise system has been in place for 25 years. there have certainly been success stories. As a regular traveller on GNER during the initial part of that 25-year period the author has a defined opinion. GNER was reportedly the only profitable element within the American group that owned it. It was obliged , by circumstance, to pass its franchise to Virgin Trains. That subsequently failed and became a national entity LNER, which it operates under today (2021).

Under the GNER banner, the breakfast in the morning and dinners on the evening train were superb. The travel was comfortable, and the trains upgraded from the diesel 125 to the 225 Electra. In other words, money was being spent, and investment was being made.

The electrification of the east coast mainline from London through to Edinburgh was a major underpinning of train journey improvement, but not initially without the occasional mishap. Powerlines down or signalling to the north of Newark or south of Stevenage areas respectively plagued an otherwise reliable service. A Land Rover vehicle falling onto the track causing a fatal accident did not help its fortunes.

Capital investment in the rail system through the two government-run entities, 'Railtrack,' which looked after the infrastructure that the trains run on, and 'Network Rail,' which coordinated the timetables and the running of the franchise-owned and operated trains within that system, was evident quite quickly after privatisation. Strictly speaking, much of the rail rolling stock was not owned by the operators but rather leased to them from central leasing companies. If the operator lost the franchise, as happened with GNER, there was no capital value to come back to the franchisee from the sale of the rolling stock; it merely morphed into the new franchise operator under a fresh lease.

Ironically the first franchise was awarded to a bus operator - Stagecoach.

Railtrack inherited a huge backlog of maintenance and dilapidation, and despite all its best efforts, was 're-nationalised' just six years after starting its venture. This was despite the rail network growing in stature, reliability, and passenger usage. (25) Could it have been that, despite the millions of pounds that franchisees were paying for the privilege of running trains that they did not own on a railway track that took their leased vehicles from one end of the British Isles to the other, the cost-benefit analysis that initiated the system was fatally flawed?

The freight moving system of rail rolling stock ownership was quite separate from the passenger services and again, timetabling was a Network Rail responsibility. Notwithstanding the ideology , both freight and passenger-carrying systems ended up as subsidiary companies of primarily European-owned, government-backed operating companies as bids from UK operators ceased to be forthcoming. There was little support from the British government, but then such support was probably against EU regulations.

The underlying problems with the main rail unions and their hostile endeavours to renationalise certain parts of the network did not assist in the smooth running of the West Coast mainline where it appears inept management and modelling errors from the Department for Transport resulted in a legal entanglement that merely cost money and achieved little. (19) Cooperation and expertise are essential components of any system's success.

How could a system that, in its first 20 years of operation improved from carrying 735 million passenger journeys to 1,660 million passenger journeys, fail financially? The Hardman & Co paper endeavours to explain. (19) It is not an easy subject; however, it is worth pursuing because this author believes that the electrification of the bus fleet has to learn from the rail experience; and not repeat it.

Under the rail franchise system, a ticket could be purchased up to three months prior to the journey with one franchise operator at a significant discount to the on-the-day ticket price and reserve a seat. A ticket could also be purchased for a through-journey, which might involve two or more franchise operators. The astute purchaser would book separate tickets with each separate franchisee-operated sector and obtain discounts on each of the tickets purchased. The sum total of the staged transaction would be markedly better value for money against an advance ticket for the total journey.

One of the *Bus Back Better* paper (5) precepts is that not only should rail journeys be interlinked and through tickets be available at the same rate as the independently acquired discount tickets outlined in the previous paragraph, but that system should also extend to the bus network. The aspiration of the government is stated that ideally, there should be an interaction between one form of travel and another such that the passenger pays for one ticket which covers the complete journey at the best price available from all elements of the routes chosen. It is a noble aspiration, but is it a practical expectation? Under the rail franchise model, the answer is an emphatic no!

The capital investments put into the rail network by Network Rail and its Railtrack predecessor cannot be denied. Both the Hardman & Co. and Adam Smith Institute acknowledge that significant financial input and improvement. However, both objective reports also state that 58% of the failings within the franchise system emanate from the centralised Network rail structure.(19)(25)

Multiple billions of pounds have been ploughed into the rail network, upgrading to full electric standards on the east to west coast mainlines and elsewhere. Railway stations have been modified and upgraded. Some were even rescued from demolition (Paddington). Those now form an unrecognisable image of their former selves. They are a pleasure to visit and pass through. They contribute to their passengers' overall enjoyable experience. That has to be the aspiration for the bus network. Passenger numbers prove that if the system is modern and attractive, the public will use it in preference to their own vehicles.

As the Hardman and Co report points out, (19) the overhead electrification equipment is hardly noticed and certainly not condemned by the travelling public. It is taken for granted. It is in place, much as the third rail on the southern region or the two power rails on the London underground are there.

The power sources are part of the infrastructure: a requirement for a safe, clean, environmentally acceptable mode of travel. What people do notice is the change to Birmingham New Street station. However, do they care about the £750 million that it cost? The answer, of course, is no.

That that £750 million will be amortised over the life of the building for 30, 40, maybe 50 years hence. The capital cost pales into insignificance if it is treated as it should be - as a capital investment over time. That is what is implicit within the bus network. The electrification of the bus network must surely take place in the most long term cost-efficient manner. If the service is good, like Birmingham Railway Station, the capital cost will not dominate the passengers' thought processes.

There is no doubt that railway 'privatisation' has yielded very clear and defined benefits that would not have happened had the system remained a nationalised entity. Competition and private funding have assisted in the process. Where the nation was involved, there were errors of judgement resulting in that stated 58% of delays and cancelled services falling at the feet of Network Rail/Railtrack. That must not happen under the newly revised Emergency Recovery Measure arrangements post-

COVID on either the railways or the bus network. It certainly does not have to.

The name change announced for Network Rail is a costly cosmetic distraction to a Quango style operation that just needs to 'up its game'. The change of name will cost £ millions for an organisation that will fundamentally fulfil the same, though enhanced, function as it has for the last 25 years. (25)

The definition of a 'long journey' will infer that a ticket will allow the right to travel to pass from one franchise holder to another or through another to a final destination. That applies to the principle outlined in the *Bus Back Better* paper from the DfT.(5) Through journeys, it submits, should be purchasable and within time limits, enabling the passenger to move, on that single ticket, from bus to train, and back to bus again, until they reach their destination. That precept is achievable with modern technology. Of course, the system must be universally applied to every passenger, not just those who carry a smartphone upon which they can rely for a QR scanner to indicate to the appropriate authority that a passenger has the right to travel.

Money will probably have to change hands on buses for a while yet, although most of the travelling public will probably use some contactless plastic card arrangement. It does seem that a significant number of the travelling public are happy to adopt the contactless card mechanism, even if it means pre-purchasing the equivalent of a London 'Oyster card' to achieve it.

Favouritism has also got to be seen to be dispelled in a system where, quite blatantly, London and the southern counties franchises have achieved marked upgrades in their rolling stock (not that they did not need to). Significant areas north of Watford are still operating what can only be described as 'ancient' rolling stock within the same network rail track system and franchise ownership of leased rolling stock to the detriment of the travelling public in those areas.

Many of those services call at Lincoln Central Station and are well-known to the author. The bus fleets of the nation follow a similar pattern. The words of the *Bus Back Better* (5) initiative to 'level up' have to result in a positive outcome for both rail and bus systems. It is bus fleet upgrades that require our focus here.

The proposal for electric bus operation, and particularly the In Motion Charging buses using a local catenary facility, would be that the local authority would be responsible for the infrastructure much as Railtrack/Network Rail [Great British Railways] are with the rail network. However, perhaps there can be joint ventures governing the three elements; electrical power supply, infrastructure to deliver power, and the vehicles that will use that power.

The use of the elements facility would be 'sold' to the bus operator operating the electric bus network, much as the electricity company (which was the local authority of yesteryear) will sell units of electricity from the generating system that produces the electricity in the first place. Commerce needs to be involved to promote development and eventual replacement at the life's end of the infrastructure involved. Where the catenary is concerned, that should be 30 to 35 years after its installation.

Roadside battery charging points will have a different life cycle, as will the overnight charging facilities envisaged in the Option 2 central garage facility for pure battery propelled vehicles involved. All of that is fundamental to the time/cost/benefit analysis that is required.

It is suggested that lack of attention to detail is probably where much of the problem with the rail franchise system has come to grief. (19) Some of the franchises were noticeably short-term and therefore not worth bidding for. Bearing in mind that the capital cost of running a system is an upfront cost that is then amortised over time, the period covered by the agreement (its term) is a crucial component.

The inconsistency in the process also involved franchises that were too long and therefore spread the recovery costs over too great a period. These aspects, and others, involved in running a transport system efficiently have to be funded on a time/cost/benefit basis coupled with the replacement cost when the underlying assets need upgrading. (26)

There is no doubt that the overseas train bidders came into the United Kingdom rail franchising system on the back of very low interest rates and their own state support as pseudo nationalised industries. We are assured that UK bus operators are aware of the shortcomings of the rail franchising system because five of their number held some of the original franchises on offer. The history of that system makes for interesting reading and begs improvement. (19)

Stagecoach is stated in the Harman & Co report as an ideal illustrative experience with rail revenues in 2011 at £525 million, reducing in early 2020 to just £2.8 million. This yielded a profit of £22.9 million in the first instance and a huge loss in respect of the second. That, too, is commerce. The share price of Stagecoach ranged from 50p up to 400p in a helter-skelter fashion over the 25 years of the franchise (which was mainly bus operation for the company) and probably says it all.

Do Deutsche Bahn, SNCF/Keolis, Trenitalia and Abellio company shares show the same tortuous variation over their period of franchise operation?

It is imperative, in light of the above, that future funding for the bus network and the stated 'levelling up' process(5) is undertaken on a mutually beneficial arrangement with local authorities providing the infrastructure and bus companies operating the system. A mutually agreed format is required covering a reasonable period of time for one to provide and the other to use that infrastructure on a commercial basis. That system should provide an environmental asset to the travelling public that they will appreciate and use. As with all commercial enterprises, the usage figures cannot be pre-empted before capital expenditure is undertaken.

What is known at the beginning of this transformation is the capital cost of each element and the expected life of that element. That information then determines just how much needs to be recovered over time within the agreement reached. If the system is more successful in operation than expected, then the reward of that success, as with the losses, stays with the operator. There is no place for sour grape politics.

The infrastructure is, and should be, a relatively fixed overhead, the benefit from which the local population gain a distinct and calculable environmental and hopefully fixed income advantage. Amortisation of the capital plus an interest rate return for the local authority is intrinsic in the calculations. The capital value of each element at the end of that time should also be a known factor. The maths is not difficult, but the quantum is. There is risk attached, and the commercial operator historically takes that risk. (19)

The gardener knows that it is a false economy to save on string when planting seeds in the garden. It is not until several weeks after sowing that the value of the string exemplifies itself by its absence when the weeds have taken over the plot, and the seedlings are hardly visible to identify. It is at that point that the string secured at each end of the row of seeds clearly displays the line of planting and thereby its value. Weeding is so much easier, and the crop is protected with the ease of removal of the offending vegetation is assured. The moral of the anecdote is: avoid false economies.

Cost can only be valued over time. The analysis extracted from the documents referred to in the foregoing chapters gives clear, unambiguous guidance. History teaches the careful observer the rest.

Chapter 22.

The Retrofit Adaptation Platform – Getting It all to Work

Just as the auto-connect unit cannot deliver its promise without the overhead catenary being in place, it also requires a complementary structure and mechanism to ensure that the auto-connecter and poles that form a part of the required structure can operate efficiently and be stowed securely when not in use.

The essence of the whole program is simplicity. In order to avoid unnecessary complications being encountered in respect of both fitting the retrofit kit and its smooth operation once installed, it should conform to a standard. A standard platform can provide a uniform facility to enable a universal application.

The power connecting poles are forced against the overhead power cables by way of two large springs at the base of each pole. Historically those poles have been physically connected, more often than otherwise guided into place by the bus crew, and similarly removed from the overhead power system. Removal is done manually or by way of a winch mechanism situated at the rear of the bus pulling down on the pole just forward of the pole head to eventually have the poles hooked under a permanent fixed hook facility situated on its own secure gantry at the rear of the roof of the vehicle. Human intervention is normally required to complete both processes.

The objective of the retrofit unit(18) is to eliminate the need for human intervention through the provision of an automated but simple process of connection and retrieval/stowage.

The principle remains the same, with a winch connected by a cord facility pulling down on the power pole towards the pole 'stowage' position in order to disconnect from the catenary and secure the poles in the stowed position. The difference is that the hook mechanism comprises a cam-operated hook-based structure that closes over the pole as it is pulled into the centre of the mechanism. Thus, the pole is secured, and the winch is automatically placed in neutral.

When the driver of the bus seeks to connect to the catenary, they select to deploy the overhead twin pole connection facility the double hook cam mechanism securing pin is retracted, and the pole retainer cam mechanism is pulled apart by the force of the springs at the base of the connecting poles. This, in turn, lifts the poles with their connecting head mechanism deployed for connection in a vertical move.

Once the connecting head meets with the overhead cable and electrical power flows to the vehicle, the power supply activates the head rotation mechanism into the in-service position, and the locating wings swivel down to allow freedom of operation during vehicle transit. The mechanism has been designed to avoid overcomplication whilst ensuring safety and security during operations.

In order to provide for the foregoing in a uniform fashion, irrespective of the hybrid bus manufacturer, and therefore roof shape variations, a universal retrofit unit can be manufactured to a standard format comprising the two interconnected elements, connection poles and retaining hook mechanisms. The common platform can then be secured to the host vehicle by way of a specifically designed adaptor bracket secured to the side of the roof at suitable mounting points.

It is envisaged that there will be three adaptor brackets on each side of the roof at appropriate points along the roofline of the host vehicle for the retrofit unit's two forward fixings to accommodate the hinge mechanism platform section for the connecting poles to be secured to. A single adaptor unit on each side toward the rear of the roofline will secure the hook capture mechanism sub-platform, retaining the pole in the 'out of service' stowage mechanism.

Each manufacturer of the respective host hybrid vehicle undergoing the retrofit process may well dictate the positioning, shape, and security mechanism for the placement of the envisaged adaptation brackets. This would be to meet the body strength requirements inherent in the host vehicle structure to support the connecting mechanism and accommodate the dimensional requirements for the standard platform to be connected by way of securing bolts.

The electrical wiring configuration should be a relatively uncomplicated matter of a design conduit to direct the power lines appropriately to streamline the fixings with the bus bodywork and/or by way of internal modifications as appropriate.

The cost of the complete retrofit unit, including the adaptor brackets and the modifications required to the host hybrid vehicle, is estimated to be circa £20,000 per vehicle at 2021 prices.

There could be a simultaneous upgrading of the interior of the vehicle. The upgrade would ideally be carried out whilst the bus is being retrofitted as another option that can be considered. Upgrading the interior might be deemed non-essential. However, it may be desirable in an aesthetic sense as well as in consideration of extending the longevity of the vehicle. The decision should perhaps be made bearing in mind that the retrofit roof unit will deliver at least another 10- to 15-years' service life to the vehicle. The interior may well be required to live up to that extension of life to fully uplift the value of the vehicle in capital as well as passenger perception terms. (1, table 4.1)

Such an upgrade is estimated to require between £15,000 and £20,000 in 2021 terms, dependent upon the vehicle age and the modification level of upgrading that is carried out. The underlying value of the vehicle would be enhanced by either or both processes in the second-hand market.

The retrofit and upgrade could be achieved within a matter of weeks. By contrast, the development and purchase of new vehicles, quite apart from the capital cost involved, would take months if not years to upgrade the fleet as is required by the urgency of the environmental situation and the number of buses involved.

The chassis life of the target vehicle is one million miles or more. The reduced vibration and general smooth running of the vehicles in 'trolleybus' mode make that assertion even more achievable.

Chapter 23.

A Summary of the Environmental Advances

The most used routes within an urban environment or city will be well known to the bus operators and passengers alike. In that respect, the number of electrically orientated buses operating on any route would also be easily assessed. The times at which 'buses' run and the passenger capacity at varying times during the day will influence the number of vehicles on the system at any one time. That necessitates careful forward planning in respect of power supply facilities irrespective of the choice of vehicle engaged with.

Working on the assumption that the buses running on the system are retrofitted hybrid diesel/electric, or electric/electric conversions to overhead power collection, what major considerations will affect the connect and disconnect profile for the routes involved? Specific routing will require to take account of the battery capacity or diesel engine running time involved in the vehicle journey between one terminus to another, which may include, say, a rural area.

A by-product of retrofitting, which has been mentioned several times in this publication, is the lack of noise and vibration otherwise associated with fossil fuel engine propelled vehicles. Judiciously planned 'In Motion Charging' bus routes can entice people back into urban/city centre communities, perhaps where households start to replace redundant shop and office facilities. The 'In Motion Charging' bus is an attractive, smooth running, quiet, vibration-free 'carrot' when re-establishing the viability of an urban centre for reinvigorating community living.

The catenary (overhead power wire system) can either be supported by posts similar to streetlamp posts or, more simplistically, by way of direct attachment into walls adjacent to the route that the 'In Motion Charging' bus will be used. The first has a cost implication for infrastructure changes to the streets, which can be quite major. The wall attachment is a far simpler and cheaper application even if a wayleave style rental is paid to the property owners, as is the case with power poles and telegraph poles in rural areas through wayleaves.

The advantage of using lamppost catenary supports is that low voltage and therefore low heat/low power consumption street light facilities can be installed concurrently with the power provision for the 'In Motion Charging' buses. Such facilities could also be used for other public service provisions where overhead wires or speakers are required, as previously mentioned.

Power substations would be required, spaced at intervals of 1.5 to 2 miles along any route to provide power to the catenary system with the appropriate switchgear incorporated to facilitate the provision of power to the buses. The number of buses operating on the system needs to be established in order to tailor the power requirement from each substation. This is to provide an optimum level that produces enough power for the number of vehicles in operation in the sector at any one time, but without 'wasting' power unnecessarily through overprovision and the unnecessary expense that involves.

There is a significant local authority interest in providing a public transportation system that is economical to operate and is silent and environmentally attuned to their requirements. The fundamental costs centre upon a broad range of variables; the vehicles being used, the infrastructure capital cost and the amortised cost of that infrastructure over time together with the attendant operational costs by way of drivers and backup support facilities are modest.

Where the actual objective of the local authority is centred upon an environmental improvement, the proposed retrofit programme is highly significant and relatively swift to implement. The relatively low cost of retrofitting an existing hybrid fleet will more than counterbalance the capital cost of the infrastructure required to operate those vehicles.

The 'In Motion Charging' bus is the only 'state-of-the-art' heavy-duty electric bus available in 2021 that meets every environmental requirement. Rolling forward for the next 10 to 15 years, the given development costs and production restrictions on the hydrogen buses (which may well provide the viable alternative to diesel over time) should marry to give a smooth succession of environmentally and cost-effective replacement scheduling. The operational costs of the trolleybus principle (even ignoring the 50% emissions target set for 2050) are more than equal to or lower than the diesel bus comparative costings and easily comparable to hydrogen and that fuel's own limitations. In Motion Charging buses are emission-free, energy-efficient and climate-friendly and would mesh well with a hybrid hydrogen-powered electric bus. (1)

Given that there is sufficient willpower to ensure an environmentally secure future (acknowledging the urgency of the requirement), there is no reason why a catenary cannot be in place within 2 to 3 years. This could coincide with the retrofitting process required for the current hybrid fleet to be fully accommodated within a substantial number of cities and large urban town environments.

With the current state-of-the-art energy recovery systems using braking power recovery, together with the vehicle standing at bus boarding points and traffic control areas, a further 25% saving of consumed energy can readily be achieved. Therefore, the cost-efficiency of retrofitting the current fleet more than offsets the amortised capital cost of installing the catenary.

Local authority budgets will benefit from the advantages of the 'In Motion Charging' bus's long life-cycle due to the power/technology efficiency exhibited by these vehicles. The public would love them. In fact, the public will be healthier due to their use, thus making other savings in terms of medical facilities focusing on asthma and other respiratory complaints. The air is instantly cleaner using the same vehicle fleet size.

A further side issue, which has been exemplified by certain hybrid vehicles, is the power required to operate efficient onboard air-conditioning/climate control systems for passenger comfort. The overhead power supply ensures that lighting, ventilation, air-conditioning, wi-fi, and other power-consuming elements within the vehicle are not compromised in an endeavour to save battery power, especially during low-speed operation when forced air is denied to the system heat exchange units for cooling is required. Power-consuming fan units have access to full power from the catenary.

Passenger comfort is known to be central to their transportation experience. That is certainly not currently the case with all hybrid vehicles, particularly rear-engined diesel vehicles on hot summers days or cold winter nights. The United Kingdom climate bears testament to the need for a more durable and flexible onboard power supply facility to be in place on slower moving urban journeys.

The innovation of the asynchronous three-phase squirrel cage-type AC induction motor, using transformed electrical supply from the overhead facility, could once again show an innovative improvement that will lead the world on its introduction into the United Kingdom. The development of supercapacitors has further enhanced the operating efficiency of the electric bus system providing, as they do, for the stability of the power supply to the vehicles. They avert 'power spike' damage.

Consideration should be given to utilising the retrofitting facilities envisaged, negating the need for staff to exit the bus to connect or disconnect the overhead poles. This is an added bonus allowing for flexibility of route and operation where connecting and disconnecting may occur on a number of locations throughout the route as the hybrid trolleybus moves from full electrical power from the overhead cables to perhaps diesel power in a suburban area or to battery power in a city-centre environment.

There is nothing currently available at an acceptable cost and development stage with a high enough production capacity and a moderate capital cost that can compete with a retrofitting program and the creation of an 'In Motion Charging' bus fleet that will work efficiently and environmentally safely over the next 10 to 30 years, including the amortised cost of the power supply infrastructure.

Older catenary systems have produced a situation where junctions demanded a restriction in vehicle speed to avoid disconnection. Modern systems will accommodate 30 mph speeds over junctions/crossovers with ease, thus giving even more strength to the argument for the installation of a catenary system with its extensive life expectancy and adaptability. The proposed catenary would avoid junctions by design to reduce stress and improve/reduce the visual impact of the system.

The installation of the system on premier routes with high-capacity passenger requirements will allow for the development of hydrogen, or other alternative transportation systems, to take place and eventually fill the appropriate gaps where the trolleybus is not appropriate or competitive. Retrofitting is a cost-effective mechanism to modernise the current environmentally unfriendly vehicles that providers can achieve in the shortest possible timescale with the best possible cost-effective price tag to achieve national environmental targets. The science is proven. The contractors capable of the retrofit/refurbishment process need to expand to meet a formidable demand envelope.

As stated in the EU TROLLEY paper, (1) "the necessary overhead contact line infrastructure puts off many potentially interested transport companies from the restructuring of their operations to a trolleybus base. However, it can be proved that the amortised investment is recovered within 20 to 25 years. The visual 'spoiling' is usually a subjective feeling, and alternative power supplies are expensive both in respect of investment and in respect of operating and maintenance costs." (1, page 47)

It is a fact that within Europe, over 30 new trolleybus systems have been put in place within the last 30 years and well over 50 new systems worldwide. The conversion of the hybrid bus fleet to be powered and replenished from overhead contact lines makes financial and environmental sense.

The versatility of the resultant vehicles will enable passenger carriage to be increased and passenger comfort to be assured. The ergonomic advantage of the 'through bus' running via a city or urban centre taking passengers from home to hospital/railhead, home to work or work to hospital/railhead or work to home from suburbia or wherever home/work is, in an optimised environmentally friendly and efficient manner is a common Government aspiration and requirement. (5) That is surely worthy of the reasonable cost involved.

As can be established from the previous chapter and by analysing the comparison of costs; there simply is no comparison other than to state that the amortised infrastructure cost. That coupled with the fact that transport companies can literally 'buy' (fund) another 10 to 15 years of vehicle use from their aging existing fleet. A modest capital outlay will put at least seven vehicles into environmentally acceptable service within the shortest timescale, compared to buying one new vehicle that achieves 14% of the environmental outcome per cost.

Humanity simply does not have the time nor the financial resource to wait for new technologies and their mass production to come online. The environmental requirement needs immediate action.

The retrofit and refurbishment option deserves urgent consideration and prompt implementation.

Chapter 24.

Beware Society's Overdependence on Electricity

From the foregoing chapters, one factor comes out above all others. There is going to be an overreliance, if not a dependence, upon electricity.

Other forms of powering vehicles, heating homes, providing light, communications, and a myriad of other services that society uses are being trashed, discarded, and time-censured. Is society going to revert to a dependence upon a single commodity that we know from our own experience following a thunderstorm can leave us extremely cold in winter due to a misplaced thunderbolt or lightning strike when the electricity power supply fails?

The extreme dependency of cryptocurrency, which is said to use vast amounts of electrical power to ensure that its binary system allows it to work for those interested in the new wave of currency, is but an example.

We must be aware as a society, not of the zealot but of the overzealous. The former is an individual (and this author may be classed as one of those). An individual can be brought to heal with reasonable ease. An overzealous society, however, can create circumstances that, in the long run, and certainly in the occurrence of a major disaster, work against that society in much the same way as the overproduction of CO_2 is said to have done. World wars have been started by such!

We surely need to exercise caution and create value for money. Ordinary money, not cryptocurrency. The metaphor may be lost, but the message is simple - options should be maintained to safeguard society as well as the planet.

On analysis the UK could be poised to "throw the baby out with the bathwater". It is a useful metaphor. How many times have you been looking for a utensil only to find the missing item in the middle of the vegetable patch following an attempt to assist by watering the garden with the washing up water?

There are two major points that arise from the analysis that has taken place within these pages.

The first is that the most time/cost-efficient way to deliver electricity to public service vehicles is by using the catenary system to slow charge them whilst they are conducting their design function in transporting people from one place to another. This is evidenced by the following:

a) In Motion Charging is efficient in the use of the electrical supply.

b) The catenary system is efficient in extending the life of the onboard batteries that are being charged.

c) The catenary system is efficient in that it does not demand high volumes of electricity at any point to the electrical supply system.

The second major point is that hybrid vehicles are capable of producing not only the motive power to ensure that buses stay in service if there is a power failure. Hybrid power also provides electricity independently to charge the onboard batteries in a way that has been well tried and tested. The technology is known to work. The downside of that system lies with the onboard batteries and the amount of power they can produce to drive the vehicle and its internal systems (heating, lighting, air conditioning and such). That feature is still awaiting the quantum leap required to make us dependent upon the battery function alone.

By utilising existing hybrid vehicles and acknowledging the advances in diesel technology, rather than damning diesel as a fuel per se, society at least holds on to the multiple-choice of power provision that is essential to its long-term survival. We must avoid over-dependency on grid generated power, which appears to be becoming all too evident.

As has been stated, manufacturing capacity has its limitations. Surely the emphasis must be placed on the replacement or upgrading of the pre-Euro VI vehicles that are in service before anything else?

That is a huge task for society and the vehicle industry to undertake. The use of new technology can progress alongside the upgrading of old technology with established and proven systems that provide environmentally acceptable levels of emissions in environments where they can be catered for adequately. That can be achieved whilst concentrating on reducing the negative effect in areas of concentrated vehicle activity producing damaging toxins, which the catenary system used for In Motion Charging of a hybrid vehicle would accommodate economically.

One hundred years ago, individual local authorities created their own electrical generating systems, which eventually became part of the National Grid. History is a great teacher. Perhaps our focus should be somewhere between individual homes and factories generating electricity for individual use via roof panels, independent wind generators or heat pumps and the national grid?

Small, nuclear-powered fusion based generators can surely be installed on brownfield sites within large conurbations to produce electricity cheaply and efficiently. They could supplement the National Grid and provide motive power to a structure that would accommodate the needs of electrically powered buses to be supplied efficiently with that power through a catenary system.

The proven economic and relatively maintenance-free catenary system will significantly improve the battery life of the vehicles that it supplies electricity to. The further advantage is the 35 years of the system's lifespan, which will be more than enough, perhaps, for a more environmentally friendly means to be found to produce hydrogen fuel than is currently the case.

It is a proven fact that catenary-powered electrical buses are the most cost-efficient and environmentally friendly public service vehicle mechanism known to man in 2021.

The argument concerning the visual impairment of overhead wires is no excuse. For those city dwellers who believe that it is, just look up from the road occasionally when you drive through the countryside. The sky is full of cables. They are there for very good reason; it is the most efficient mechanism for those cables and the electricity they carry to go from A to B. Time has proven their worth.

As this book shows, we have had 150 years of development of public service vehicles that have shown us the remarkable advances that have been made from the hybrid vehicles of the 1890s to the modern manifestation of that same technology today. History also condemns the authorities for their overconfidence in diesel power and the consequential decimation of the trolleybus network in the United Kingdom, which has exacerbated the current dichotomy. Thank goodness there are still 44,000 examples of trolleybuses running worldwide, evidencing the error of that decision. Clean power and silent running. However, there has to be a compromise to achieve the most practical solution to the environmentally targeted analysis and consequential actions.

We can learn something from that historic act of misjudgement. The hybrid bus is perhaps an essential component of a solution.

Should society, through those in authority, capitalise on that learning process and combine the two technologies to satisfy the needs of the travelling public? Surely the marriage of a highly efficient catenary system to provide In Motion Charging and propellant power to the vehicle in an urban environment and the Euro VI environmentally acceptable diesel engine power for suburban and rural routes, where nature provides an environmental balance, can be acceptable?

The time/cost/benefit analyses are very clear. Facts must prevail over Luddite emotions. The next century's history starts here. It will do well to dwell on the achievements of the last century in its decision-making process. Hopefully, the facts within these covers will aid that process.

Appendix

Text Listing for Reference Material and Learned Papers

1. EU Take-up guide for the Replacement of urban Diesel buses by Trolleybuses2013 http://www.trolley-project.eu/fileadmin/user_upload/Outputs/TROLLEY_PR6_Take_up_Guide_Core_Output_4.3.8_Annex_146_PP05_LVB.pdf

2. *USA Costs and Benefits of Electrifying and automating Bus Transit Fleets* Kockelman / Mohamed Feb 2020 https://res.mdpi.com/d_attachment/sustainability/sustainability-12-03977/article_deploy/sustainability-12-03977.pdf

3. Steve Swain managing director of Chartwell Bus Sales Ltd Bus market data - second-hand vehicles. https://chartwellbussales.co.uk

4. *'Trolleybus'* reviews of worldwide trolleybus data. Dave Chick editor. http://www.britishtrolleybus.org.uk

5. DfT (2021) *Bus Back Better* https://assets.publishing.service.gov.uk

6. DfT (2016) *Value for money assessment for major bus-related schemes.* https://www.gov.uk/government/publications/value-for-money-of-major-bus-related-schemes

7. DfT Table BUS 0602 Number of buses used as PSV by metropolitan area status

8. DfT Table BUS 0609a Percentage of buses used as PSV by emission standards
9. DfT Table BUS 0609b Percentage of buses used as PSV by fuel type
10. The Electric Bus Group - 14/4/2021 Trolleybus UK Electric vehicle charging options http://www.trolleybus.co.uk
11. 2016 Leeds Powertrain proposal - public enquiry findings
12. Manchester Tram cost-benefit
13. Sheffield Tram cost-benefit
14. The recipe for increased passenger interest report
15. Birmingham battery tramway
16. Multibus™ Connector head unit
17. Multibus™ Pole stowage cam hook unit
18. Multibus™ Proposed standard retrofit platform - pole connectors
19. *The Failure of the Rail Franchises.* Nigel Hawkins, Harman & Co
20. DfT table vehicles licenced in Great Britain
21. Wright bus production by new vehicle registrations 2015-2020
22. Toronto Transport Commission [TTC commission green bus technology plan update.pdf]

23. Vancouver Transit Operator 'Translink' [Translink agenda mayors council public mtg.pdf]

24. The Trolleybus Museum Sandtoft, North Lincolnshire

25. Rail Franchise article Matthew Lesh, The Adam Smith Institute. Sunday Telegraph 23/5/2021

26. Railfuture.org.uk/article1796-Franchise-failure

27. *It's not the bus. The issue is the infrastructure.* Joe Bamford

Author's Credentials

Terence P. O'Halloran was born in June 1944.

In January 1960, Terry joined the Royal Air Force as an engineering apprentice at No 1 School of Technical Training RAF Halton and served as an engineer for 12 years. In 1971, he became a life assurance and financial planner.

In 1974, Terence started writing books as sales guides for those in the life insurance and financial planning fraternity. In 1977, he set up Life Publications Limited to publish his own titles, *A Look at Life Assurance* and *First in the Indian Skies*, which was printed in 1981 (the history of the 31 Sqn RAF). In 1977 he achieved an ONC in Business Studies at Lincoln Technical College along with an Associateship of the Chartered Insurance Institute by examination.

In 1978, Terence was appointed by Witherby and Co., London publishers, to write books for them to publish: *You Sign the Little Cheque and We Sign the Big One*, *You Sign*, and *Mountains Out of Molehills* together with *Trusts: a Practical Guide* and others. On 15th June 1990, Terence was awarded a Batchelor of Science degree in Business Administration, and on 30th March 2000, he achieved a Fellowship of the Chartered Insurance Institute by examination.

In 2003, Wetherby's closed their financial services publishing department. Thereafter Life Publications Ltd produced *'Trusts' A Practical Guide (2nd edition)* in 2000. Terence designed and published *Parents' Poetry* by Judy Theobald to celebrate 25 years in business. In February 2006, he qualified as a Chartered Financial Planner by examination.

June 2011 heralded the publication of *Hindsight - The Foresight Saga* as a retirement gift to his clients. Other books include: *If Only Politicians Had Brains*, published in 1999; *Building A Business on Bacon and Eggs* in 2003 and *The Fight for Our Post Office* in 2008.

Outside of a successful career spanning 44 years, Terence completed 13 years in the Gainsborough Junior Chamber of Commerce (Jaycees International) as a President and International Senator. He was also President of the Lincolnshire Chamber of Commerce for two terms of office.

In 1974, he was a founder member and Branch Chairman of the National Federation of Self Employed. That organisation became the Federation of Small Businesses, of which he was on the National Executive Committee and their national pension spokesman until 2003. He was also a founder member of the British Insurance Brokers Association. Terence spent five years as Executive and Vice President of the Life Insurance Association and three years as the managing editor of Prospect Magazine - the LIA's monthly mouthpiece.

After leaving Jaycees at age 40, Terence began his 35 years membership of Bailgate Rotary Club, ceasing in 2018.

That including two Presidential terms. He was four times chairman/organiser of Rotary District 1270 Disabled Games 1998/1999 and 2004/2005. In recognition of this, he received the Paul Harris award.

Over his working life, several products have been designed and implemented with good effect.

A busy life spanning several decades has witnessed the creation and organisation of various 'study groups.' They were formed and became successful mutual learning bases for Independent Financial Advisers, also, company salespeople, aspiring public speakers and professionals, plus groups of multi-disciplined businesspeople at several breakfast and luncheon clubs. Learning became a pleasurable experience from peer group resources. The creative pattern for the study group principles was encapsulated in his book Building a Business on Bacon and Eggs, written in 2003.

In 2019, Multibus Limited was formed to explore and create, as appropriate, the reintroduction of In Motion Charging of electric-powered public service (and ultimately other large) vehicles. This was to be done through the retrofitting of predominantly diesel-engined hybrid buses to accept electrical power from overhead cables using an auto-connect and disconnect system that precludes physical intervention. The initial development has been achieved, and now, in late 2021, it requires adoption, implementation and funding. It is finally time for the 'trolleybus revolution' in the UK and wider markets.

Made in the USA
Monee, IL
12 September 2026

62975833R00148